ENVIRONMENTAL RESEARCH ADVANCES

CLEAN AIR AND WATER

SUMMARIES OF MAJOR FEDERAL LAWS

ENVIRONMENTAL RESEARCH ADVANCES

Additional books in this series can be found on Nova's website under the Series tab.

Additional E-books in this series can be found on Nova's website under the E-book tab.

ENVIRONMENTAL RESEARCH ADVANCES

CLEAN AIR AND WATER

SUMMARIES OF MAJOR FEDERAL LAWS

CAROLINE MCNULTY

AND

ROGER HUBER

EDITORS

New York

For permission to use material from this book please contact us:
Telephone 631-231-7269; Fax 631-231-8175
Web Site: http://www.novapublishers.com

LIBRARY OF CONGRESS CATALOGING-IN-PUBLICATION DATA

ISBN: 978-1-62808-444-3

Published by Nova Science Publishers, Inc. † New York

CONTENTS

PREFACE

The principal statute addressing air quality concerns, the Clean Air Act was first enacted in 1955, with major revisions in 1970, 1977, and 1990. The Act requires the EPA to set health-based standards for ambient air quality, set deadlines for the achievement of those standards by state and local governments, and requires the EPA to set national emission standards for large or ubiquitous sources of air pollution, including motor vehicles, power plants, and other industrial sources. The Safe Drinking Water Act (SDWA) is the key federal law for protecting public water supplies from harmful contaminants. First enacted in 1974 and substantially amended in 1986 and 1996, the act is administered through programs that establish standards and treatment requirements for public water supplies, control underground injection of wastes, finance infrastructure projects, and protect sources of drinking water. This book describes both Act's major provisions and provides tables listing all major amendments, with the year of enactment and Public Law number, and cross-referencing sections of the Acts with the major U.S. Code sections of the codified statute.

Chater 1 – The principal statute addressing air quality concerns, the Clean Air Act was first enacted in 1955, with major revisions in 1970, 1977, and 1990. The Act requires EPA to set health-based standards for ambient air quality, sets deadlines for the achievement of those standards by state and local governments, and requires EPA to set national emission standards for large or ubiquitous sources of air pollution, including motor vehicles, power plants, and other industrial sources. In addition, the Act mandates emission controls for sources of 187 hazardous air pollutants, establishes a capand-trade program to limit acid rain, requires the prevention of significant deterioration of air quality in areas with clean air, requires a program to restore visibility impaired by regional haze in national parks and wilderness areas, and implements the Montreal Protocol to phase out most ozone-depleting chemicals.

This report describes the Act's major provisions and provides tables listing all major amendments, with the year of enactment and Public Law number, and cross-referencing sections of the Act with the major U.S. Code sections of the codified statute.

Chater 2 – This report summarizes the Safe Drinking Water Act (SDWA) and its major programs and regulatory requirements. This report includes the drinking water security provisions added to the SDWA by the Public Health Security and Bioterrorism Preparedness and Response Act of 2002 (P.L. 107-188).

The SDWA, Title XIV of the Public Health Service Act, is the key federal law for protecting public water supplies from harmful contaminants. First enacted in 1974 and substantially amended in 1986 and 1996, the act is administered through programs that establish standards and treatment requirements for public water supplies, control underground injection of wastes, finance infrastructure projects, and protect sources of drinking water. The 1974 law established the current federal-state arrangement in which states may be delegated primary implementation and enforcement authority for the drinking water program. The state-administered Public Water Supply Supervision (PWSS) Program remains the basic program for regulating the nation's public water systems, and 49 states have assumed this authority.

The last major reauthorization of the act was done through the Safe Drinking Water Act Amendments of 1996 (P.L. 104-182), which generally authorized appropriations for SDWA programs through FY2003. As with other EPA-administered statutes having expired funding authority, Congress has continued to appropriate funds for the ongoing SDWA programs.

In addition to reviewing key programs and requirements of the SDWA, this report includes statistics on the number and types of regulated public water systems. It also provides tables that list all major amendments, with the year of enactment and public law number, and that cross- reference sections of the act with the major *U.S. Code* sections of the codified statute.

Chater 3 – The principal law governing pollution of the nation's surface waters is the Federal Water Pollution Control Act, or Clean Water Act. Originally enacted in 1948, it was totally revised by amendments in 1972 that gave the act its current dimensions. The 1972 legislation spelled out ambitious programs for water quality improvement that have since been expanded and are still being implemented by industries and municipalities.

This report presents a summary of the law, describing the statute without discussing its implementation.

The Clean Water Act consists of two major parts, one being the provisions which authorize federal financial assistance for municipal sewage treatment plant construction. The other is the regulatory requirements that apply to industrial and municipal dischargers. The act has been termed a technology-forcing statute because of the rigorous demands placed on those who are regulated by it to achieve higher and higher levels of pollution abatement under deadlines specified in the law. Early on, emphasis was on controlling discharges of conventional pollutants (e.g., suspended solids or bacteria that are biodegradable and occur naturally in the aquatic environment), while control of toxic pollutant discharges has been a key focus of water quality programs more recently.

Prior to 1987, programs were primarily directed at point source pollution, that is, wastes discharged from discrete sources such as pipes and outfalls. Amendments to the law in that year authorized measures to address nonpoint source pollution (runoff from farm lands, forests, construction sites, and urban areas), which is estimated to represent more than 50% of the nation's remaining water pollution problems. The act also prohibits discharge of oil and hazardous substances into U.S. waters.

Under this act, federal jurisdiction is broad, particularly regarding establishment of national standards or effluent limitations. Certain responsibilities are delegated to the states, and the act embodies a philosophy of federal-state partnership in which the federal government sets the agenda and standards for pollution abatement, while states carry out day-to-day activities of implementation and enforcement.

To achieve its objectives, the act is based on the concept that all discharges into the nation's waters are unlawful, unless specifically authorized by a permit, which is the act's principal enforcement tool. The law has civil, criminal, and administrative enforcement provisions and also permits citizen suit enforcement.

Financial assistance for constructing municipal sewage treatment plants and certain other types of water quality improvements projects is authorized under title VI. It authorizes grants to capitalize State Water Pollution Control Revolving Funds, or loan programs. States contribute matching funds, and under the revolving loan fund concept, monies used for wastewater treatment construction are repaid to states, to be available for future construction in other communities.

Chater 4 – The Marine Protection, Research, and Sanctuaries Act has two basic aims: to regulate intentional ocean disposal of materials, and to authorize related research. Permit and enforcement provisions of the law are often referred to as the Ocean Dumping Act. The basic provisions of the act have remained virtually unchanged since 1972, when it was enacted to establish a comprehensive waste management system to regulate disposal or dumping of all materials into marine waters that are within U.S. jurisdiction, although a number of new authorities have been added. This report presents a summary of the law.

In: Clean Air and Water
Editors: C. McNulty and R. Huber
ISBN: 978-1-62808-444-3

Chapter 1

CLEAN AIR ACT: A SUMMARY OF THE ACT AND ITS MAJOR REQUIREMENTS*

James E. McCarthy, Claudia Copeland and Linda-Jo Schierow

SUMMARY

The principal statute addressing air quality concerns, the Clean Air Act was first enacted in 1955, with major revisions in 1970, 1977, and 1990. The Act requires EPA to set health-based standards for ambient air quality, sets deadlines for the achievement of those standards by state and local governments, and requires EPA to set national emission standards for large or ubiquitous sources of air pollution, including motor vehicles, power plants, and other industrial sources. In addition, the Act mandates emission controls for sources of 187 hazardous air pollutants, establishes a capand-trade program to limit acid rain, requires the prevention of significant deterioration of air quality in areas with clean air, requires a program to restore visibility impaired by regional haze in national parks and wilderness areas, and implements the Montreal Protocol to phase out most ozone-depleting chemicals.

This report describes the Act's major provisions and provides tables listing all major amendments, with the year of enactment and Public Law

---

* This is an edited, reformatted and augmented version of a Congressional Research Service publication, CRS Report for Congress RL30853, from www.crs.gov, prepared for Members and Committees of Congress, dated January 11, 2013.

number, and cross-referencing sections of the Act with the major U.S. Code sections of the codified statute.

INTRODUCTION

The authorities and responsibilities of the Environmental Protection Agency (EPA) derive primarily from a dozen major environmental statutes. This report provides a concise summary of one of those statutes, the Clean Air Act. It provides a very brief history of federal involvement in air quality regulation and of the provisions added by legislation in 1970, 1977, and 1990; it explains major authorities contained in the Act; it defines key terms; and it lists references for more detailed information on the Act and its implementation.

While this report attempts to present the essence of the Act, it is necessarily incomplete. Many details and secondary provisions are omitted. In addition, the report describes the statute largely without discussing its implementation. Statutory deadlines to control emissions and achieve particular mandates have often been missed as a result of delayed standard-setting by EPA, delayed action on implementation by states and local governments, or law suits brought by interested parties.

OVERVIEW

The Clean Air Act, codified as 42 U.S.C. 7401 *et seq.*, seeks to protect human health and the environment from emissions that pollute ambient, or outdoor, air. It requires the Environmental Protection Agency to establish minimum national standards for air quality, and assigns primary responsibility to the states to assure compliance with the standards. Areas not meeting the standards, referred to as "nonattainment areas," are required to implement specified air pollution control measures. The Act establishes federal standards for mobile sources of air pollution and their fuels and for sources of 187 hazardous air pollutants, and it establishes a cap-and-trade program for the emissions that cause acid rain. It establishes a comprehensive permit system for all major sources of air pollution. It also addresses the prevention of pollution in areas with clean air and protection of the stratospheric ozone layer.

Like many other programs administered by the Environmental Protection Agency, federal efforts to control air pollution have gone through several phases, beginning with information collection, research, and technical assistance, before being strengthened to establish federal standards and enforcement.

Table 1. Clean Air Act and Amendments (codified generally as 42 U.S.C. 7401-7671)

Year	Act	Public Law Number
1955	Air Pollution Control Act	P.L. 84-159
1959	Reauthorization	P.L. 86-353
1960	Motor vehicle exhaust study	P.L. 86-493
1963	Clean Air Act Amendments	P.L. 88-206
1965	Motor Vehicle Air Pollution Control Act	P.L. 89-272, Title I
1966	Clean Air Act Amendments of 1966	P.L. 89-675
1967	Air Quality Act of 1967 National Air Emission Standards Act	P.L. 90-148
1970	Clean Air Act Amendments of 1970	P.L. 91-604
1973	Reauthorization	P.L. 93-13
1974	Energy Supply and Environmental Coordination Act of 1974	P.L. 93-319
1977	Clean Air Act Amendments of 1977	P.L. 95-95
1980	Acid Precipitation Act of 1980	P.L. 96-294, Title VII
1981	Steel Industry Compliance Extension Act of 1981	P.L. 97-23
1987	Clean Air Act 8-month Extension	P.L. 100-202
1990	Clean Air Act Amendments of 1990	P.L. 101-549
1991	Technical correction to list of hazardous air pollutants	P.L. 102-187
1995-96	Relatively minor laws amending the Act	P.L. 104-6, 59, 70, 260
1998	Amended § 604 re methyl bromide	P.L. 105-277, Section 764
1998	Border Smog Reduction Act of 1998	P.L. 105-286
1999	Chemical Safety Information, Site Security and Fuels Regulatory Relief Act	P.L. 106-40
2004	Amendments to §209 re small engines	P.L. 108-199, Division G, Title IV, Section 428
2005	Energy Policy Act of 2005 (amended §211 re fuels)	P.L. 109-58
2007	Energy Independence and Security Act of 2007 (amended §211 re fuels)	P.L. 110-140

Federal legislation addressing air pollution was first passed in 1955, prior to which air pollution was the exclusive responsibility of state and local levels of government.The federal role was strengthened in subsequent amendments, notably the Clean Air Act Amendments of 1970, 1977, and 1990. The 1970 amendments established the procedures under which EPA sets national standards for ambient air quality, required a 90% reduction in emissions from new automobiles by 1975, established a program to require the best available control technology at major new sources of air pollution, established a program to regulate air toxics, and greatly strengthened federal enforcement authority. The 1977 amendments adjusted the auto emission standards, extended deadlines for the attainment of air quality standards, and added the Prevention of Significant Deterioration program to protect air cleaner than national standards.

Changes to the Act in 1990 included provisions to (1) classify most nonattainment areas according to the extent to which they exceed the standard, tailoring deadlines, planning, and controls to each area's status; (2) tighten auto and other mobile source emission standards; (3) require reformulated and alternative fuels in the most polluted areas; (4) revise the air toxics section, establishing a new program of technology-based standards and addressing the problem of sudden, catastrophic releases of air toxics; (5) establish an acid rain control program, with a marketable allowance scheme to provide flexibility in implementation; (6) require a state-run permit program for the operation of major sources of air pollutants; (7) implement the Montreal Protocol to phase out most ozone-depleting chemicals; and (8) update the enforcement provisions so that they parallel those in other pollution control acts, including authority for EPA to assess administrative penalties.

The 1990 amendments also authorized appropriations for clean air programs through FY1998. The Act has not been reauthorized since then. House rules require enactment of an authorization before an appropriation bill can be considered; but this requirement can be waived and frequently has been. Thus, while authorization of appropriations in the Clean Air Act (and most other environmental statutes) has expired, programs have continued and have been funded. The Act's other legal authorities, to issue and enforce regulations, are, for the most part, permanent and are not affected by the lack of authorization.

The remainder of this report describes major programs required by the Act, with an emphasis on the changes established by the 1990 amendments.

NATIONAL AMBIENT AIR QUALITY STANDARDS

In section 109, the Act requires EPA to establish National Ambient Air Quality Standards (NAAQS) for air pollutants that endanger public health or welfare,[1] in the Administrator's judgment, and whose presence in ambient air results from numerous or diverse sources. The NAAQS must be designed to protect public health with an adequate margin of safety and to protect public welfare from any known or anticipated adverse effects. Using this authority, EPA has promulgated NAAQS for six air pollutants: sulfur dioxide (SO_2), particulate matter ($PM_{2.5}$ and PM_{10}), nitrogen dioxide (NO_2), carbon monoxide (CO), ozone,[2] and lead. The Act requires EPA to review the scientific data upon which the standards are based every five years, and revise the standards, if necessary.

More often than not, EPA has taken more than five years in reviewing the standards, but the establishment of a deadline has allowed interested parties to force review of the standards by filing suit.

Originally, the Act required that the NAAQS be attained by 1977 at the latest, but the states experienced widespread difficulty in complying with this deadline.

As a result, the deadlines for achieving NAAQS have been extended several times. Under the 1990 amendments, most areas not in attainment with NAAQS must meet special compliance schedules, staggered according to the severity of an area's air pollution problem. The amendments also established specific requirements for each nonattainment category, as described below.

STATE IMPLEMENTATION PLANS

While the Act authorizes the EPA to set NAAQS, the states are responsible for establishing procedures to attain and maintain the standards. Under Section 110 of the Act, the states adopt plans, known as State Implementation Plans (SIPs), and submit them to EPA to ensure that they are adequate to meet statutory requirements.

SIPs are based on emission inventories and computer models to determine whether air quality violations will occur. If these data show that standards would be exceeded, the state must impose additional controls on existing sources to ensure that emissions do not cause "exceedances" of the standards. Proposed new and modified sources must obtain state construction permits in which the applicant shows how the anticipated emissions will not exceed allowable limits. In nonattainment areas, emissions from new or modified sources must also be offset by reductions in emissions from existing sources.

The 1990 amendments require EPA to impose sanctions in areas which fail to submit a SIP, fail to submit an adequate SIP, or fail to implement a SIP: unless the state corrects such failures, a 2-to-1 emissions offset for the construction of new polluting sources is imposed 18 months after notification to the state, and a ban on most new federal highway grants is imposed six months later.

An additional ban on air quality grants is discretionary. Ultimately, a Federal Implementation Plan may be imposed if the state fails to submit or implement an adequate SIP.

The amendments also require that, in nonattainment areas, no federal permits or financial assistance may be granted for activities that do not "conform" to a State Implementation Plan. This requirement can cause a temporary suspension in funding for most new highway and transit projects if an area fails to demonstrate that the emissions caused by such projects are consistent with attainment and maintenance of ambient air quality standards. Demonstrating conformity of transportation plans and SIPs is required in nonattainment areas whenever new plans are submitted.

NONATTAINMENT REQUIREMENTS

In a major departure from the prior law, the 1990 Clean Air Act Amendments grouped most nonattainment areas into classifications based on the extent to which the NAAQS was exceeded, and established specific pollution controls and attainment dates for each classification. These requirements are described here as spelled out in Sections 181-193 of the Act.[3]

Nonattainment areas are classified on the basis of a "design value," which is derived from the pollutant concentration (in parts per million or micrograms per cubic meter) recorded by air quality monitoring devices. The design value for the one-hour ozone standard was the fourth highest hourly reading measured during the most recent three-year period. Using these design values, the Act created five classes of ozone nonattainment, as shown in *Table 2*. Only Los Angeles fell into the "extreme" class, but 97 other areas were classified in one of the other four ozone categories.

The classification system and design values have since been adapted twice as the ozone standard has been revised. Under the current (2008) standard, there are 46 nonattainment areas, all but 10 of which are classified as "marginal" (see *Table 3*).

A simpler classification system established moderate and serious nonattainment areas for carbon monoxide and particulate matter with correspondingly more stringent control requirements for the more polluted class.

Table 2. Statutory Ozone Nonattainment Classifications

Class	Marginal	Moderate	Serious	Severe	Extreme
Deadline	1993	1996	1999	2005-2007[a]	2010
Areas[b]	42 areas	32 areas	14 areas	9 areas	1 area
Design Value	0.121 ppm-0.138 ppm	0.138 ppm-0.160 ppm	0.160 ppm-0.180 ppm	0.180 ppm-0.280 ppm	>0.280 ppm

[a] Areas with a 1988 design value between 0.190 and 0.280 ppm had 17 years to attain; others had 15 years.

[b] Number of areas in each category as of the date of enactment.

As shown in *Table 2*, the statutory attainment deadlines for ozone nonattainment areas stretched from 1993 to 2010, depending on the severity of the problem. Under the current eight-hour ozone standard, shown in *Table 3*, these deadlines are changed to 2015 to 2032.

Table 3. Current Ozone Nonattainment Classifications

Class	Marginal	Moderate	Serious	Severe	Extreme
Deadline	2015	2018	2021	2027-2029[a]	2032
Areas[b]	36 areas	3 areas	2 areas	3 areas	2 areas
Design Value	0.076 ppm-0.086 ppm	0.086 ppm-0.100ppm	0.100 ppm-0.113ppm	0.113 ppm-0.175 ppm	>0.175 ppm

[a] Areas with a design value between 0.119 and 0.175 ppm have 17 years to attain; others have 15 years.

[b] Number of areas in each category as of the date of designation (May 2012).

For carbon monoxide, the attainment date for moderate areas was December 31, 1995, and for serious areas, December 31, 2000. For particulate matter, the deadline for areas designated moderate nonattainment as of 1990 was December 31, 1994; for those areas subsequently designated as moderate, the deadline is six years after designation. For serious areas, the respective deadlines are December 31, 2001, or 10 years after designation.

Requirements for Ozone Nonattainment Areas

Although areas with more severe air pollution problems have a longer time to meet the standards, more stringent control requirements are imposed in areas with worse pollution.

A summary of the primary ozone control requirements for each nonattainment category follows.

Marginal Areas

- Inventory emissions sources (to be updated every three years).
- Require 1.1 to 1 offsets (i.e., new major emission sources of volatile organic compounds [VOCs] must reduce VOC emissions from existing facilities in the area by 10% more than the emissions of the new facility).
- Impose reasonably available control technology (RACT) on all major sources emitting more than 100 tons per year for the nine industrial categories where EPA had already issued control technique guidelines describing RACT prior to 1990.

Moderate Areas

- Meet all requirements for marginal areas.
- Impose a 15% reduction in VOC emissions in six years.
- Adopt a basic vehicle inspection and maintenance program.
- Impose RACT on all major sources emitting more than 100 tons per year for all additional industrial categories where EPA will issue control technique guidelines describing RACT.
- Require vapor recovery at gas stations selling more than 10,000 gallons per month.
- Require 1.15 to 1 offsets.

Serious Areas

- Meet all requirements for moderate areas.
- Reduce definition of a major source of VOCs from emissions of 100 tons per year to 50 tons per year for the purpose of imposing RACT.
- Reduce VOCs 3% annually for years 7 to 9 after the 15% reduction already required by year 6.
- Improve monitoring.
- Adopt an enhanced vehicle inspection and maintenance program.

- Require fleet vehicles to use clean alternative fuels.
- Adopt transportation control measures if the number of vehicle miles traveled in the area is greater than expected.
- Require 1.2 to 1 offsets.
- Adopt contingency measures if the area does not meet required VOC reductions.

Severe Areas

- Meet all requirements for serious areas.
- Reduce definition of a major source of VOCs from emissions of 50 tons per year to 25 tons per year for the purpose of imposing RACT.
- Adopt specified transportation control measures.
- Implement a reformulated gasoline program.
- Require 1.3 to 1 offsets.
- Impose $5,000 per ton penalties on major sources if the area does not meet required reductions.

Extreme Areas

- Meet all requirements for severe areas.
- Reduce definition of a major source of VOCs from emissions of 25 tons per year to 10 tons per year for the purpose of imposing RACT.
- Require clean fuels or advanced control technology for boilers emitting more than 25 tons per year of NO_x.
- Require 1.5 to 1 offsets.

Requirements for Carbon Monoxide Nonattainment Areas

As with ozone nonattainment areas, carbon monoxide (CO) nonattainment areas are subjected to specified control requirements, with more stringent requirements in Serious nonattainment areas. A summary of the primary CO control requirements for each nonattainment category follows.

Moderate Areas

- Conduct an inventory of emissions sources.
- Forecast total vehicle miles traveled in the area.
- Adopt an enhanced vehicle inspection and maintenance program.
- Demonstrate annual improvements sufficient to attain the standard.

Serious Areas

- Adopt specified transportation control measures.
- Implement an oxygenated fuels program for all vehicles in the area.
- Reduce definition of a major source of CO from emissions of 100 tons per year to 50 tons per year if stationary sources contribute significantly to the CO problem.

Serious areas failing to attain the standard by the deadline had to revise their SIP and demonstrate reductions of 5% per year until the standard was attained.

All areas have now attained the standard.

Requirements for Particulate Nonattainment Areas

Particulate (PM_{10}) nonattainment areas are also subject to specified control requirements. These are:

Moderate Areas

- Require permits for new and modified major stationary sources of PM_{10}.
- Impose reasonably available control measures (RACM).

Serious Areas

- Impose best available control measures (BACM).
- Reduce definition of a major source of PM_{10} from 100 tons per year to 70 tons per year.

In July 1997, EPA promulgated new standards for fine particulates ($PM_{2.5}$). Implementation of the $PM_{2.5}$ standards was delayed by court challenges and by the initial absence of a monitoring network capable of measuring the pollutant.

Nonattainment areas for $PM_{2.5}$ were designated on April 14, 2005. States had three years subsequent to designation to submit State Implementation Plans.

Revisions to the NAAQS promulgated in October 2006 strengthened the $PM_{2.5}$ standard and triggered a new round of nonattainment area designations.

TRANSPORTED AIR POLLUTION

Meeting the nation's clean air standards can be complicated, as air pollution is no respecter of political boundaries or subdivisions. This problem of transported air pollutants has come into particular focus as states and EPA attempt to develop effective compliance strategies to achieve both the ozone and the $PM_{2.5}$ NAAQS. Under Section 110(a)(2)(D), SIPs must include adequate provisions to prevent sources within that state from contributing significantly to nonattainment in one or more downwind states.

If EPA finds a SIP inadequate to achieve a NAAQS, it must require the affected state to submit a revised SIP that includes sufficient measures to bring that state into compliance.

This is known as a "SIP Call." The 1990 Clean Air Act amendments provided EPA and the states with new tools to address the transport problem through this provision. One of those tools is Section 176A, a provision that permits the EPA, either on its own or by petition from any state, to establish a transport region to address regional pollution problems contributing to violations of a primary NAAQS. A commission of EPA and state officials is constituted to make recommendations to EPA on appropriate mitigation strategies. Based on the commission's findings and recommendations, EPA is then required under section 110(k)(5) to notify affected states of inadequacies in their current state implementation plans and to establish deadlines (not to exceed 18 months) for submitting necessary revisions. Besides authorizing administratively-created transport regions, the 1990 amendments statutorily created an Ozone Transport Region (OTR) in the Northeast. This provision (Section 184 of the act) required specific additional controls for all areas (not only nonattainment areas) in that region, and established the Ozone Transport Commission for the purpose of recommending to EPA regionwide controls affecting all areas in the region.

The transport issue may also be addressed by affected downwind states through a Section 126 petition. As amended by the 1990 Clean Air Act amendments, under Section 126(b) any state or political subdivision may petition EPA for a finding that a major source or group of stationary sources located in another state is emitting pollutants that "significantly contribute" to the nonattainment of a NAAQS by their state. EPA is to respond to the petition within 60 days.

If the petition is granted, the offending sources must cease operations within three months unless the sources comply with emission controls and the compliance schedules as determined by EPA to bring them into compliance with the section. Section 126 has rarely been used, although it has proven useful to EPA in some cases as backup authority where there might be challenges to a SIP call.

EMISSION STANDARDS FOR MOBILE SOURCES

Title II of the Clean Air Act has required emission standards for automobiles since 1968. The 1990 amendments significantly tightened these standards: for cars, the hydrocarbon standard was reduced by 40% and the nitrogen oxides (NO_x) standard by 50%. These standards—referred to as "Tier 1" standards—were phased in over the 1994-1996 model years.

The amendments envisioned a further set of reductions ("Tier 2" standards), but not before model year 2004. For Tier 2 standards to be promulgated, the agency was first required to report to Congress concerning the need for further emission reductions, the availability of technology to achieve such reductions, and the cost-effectiveness of such controls compared to other means of attaining air quality standards. EPA submitted this report to Congress in August 1998, concluding that further emission reductions were needed and that technology to achieve such reductions was available and cost-effective. Tier 2 standards, requiring emission reductions of 77% to 95% from cars and light trucks were promulgated in February 2000, and were phased in over the 2004-2009 model years. To facilitate the use of more effective emission controls, the standards also required a more than 90% reduction in the sulfur content of gasoline, beginning in 2004.

The 1990 amendments also required that oxygenated gasoline, designed to reduce emissions of carbon monoxide, be sold in the worst CO nonattainment areas and that "reformulated" gasoline (RFG), designed to reduce emissions of volatile organic compounds and toxic air pollutants, be sold in the nine worst ozone nonattainment areas (Los Angeles, San Diego, Houston, Baltimore, Philadelphia, New York, Hartford, Chicago, and Milwaukee); metropolitan Washington, DC; and four areas in California were added to the mandatory list later. Other ozone nonattainment areas can opt in to the RFG program; as of 2006, additional areas in 11 states had done so.

The fuels provisions were modified by the Energy Policy Act of 2005 (EPACT), removing the requirement that RFG contain oxygenates. Instead, EPACT required the use of increasing amounts of renewable fuel, most likely to be ethanol, in motor fuels, beginning in 2006. The Energy Independence and Security Act of 2007 further strengthened the renewable fuel requirements.

Use of alternative fuels and development of cleaner engines was also to be stimulated by the Clean-Fuel Fleet Program. In all of the most seriously polluted ozone and CO nonattainment areas, centrally fueled fleets of 10 or more passenger cars and light-duty trucks must purchase at least 30% clean-fuel vehicles when they add new vehicles to existing fleets, starting in 1999. (The Act originally required the program to begin in 1998, but the start was delayed by a year.) The percentage rose to 50% in 2000 and 70% in 2001. Heavy-duty fleets are required to purchase at least 50% clean-fuel vehicles annually. A clean fuel vehicle is one which meets Low Emission Vehicle (LEV) standards and operates on reformulated gasoline, reformulated diesel, methanol, ethanol, natural gas, liquefied petroleum gas, hydrogen, or electricity.

In addition to the above program, California's Zero Emission Vehicle (ZEV) program also is intended to promote the development of alternative fuels and vehicles. Section 209(b) of the Clean Air Act allows the EPA Administrator to grant California the authority to develop its own vehicle emissions standards if those standards are at least as stringent as the federal standards and if the state demonstrates that it needs the standards to meet compelling and extraordinary conditions. In addition to setting more stringent standards for all vehicles, California used this authority to establish a program requiring auto manufacturers to sell ZEVs (electric or hydrogen fuel cell vehicles) in the state beginning in 2003. This program has been substantially modified since it was enacted, and now allows credit for hybrid and partial ZEV vehicles in addition to true ZEVs, but it has served as an incubator for lower emission technologies since its adoption. Section 177 of the Act allows other states to adopt California's stricter standards: at least ten states (Connecticut, Maine, Massachusetts, New Jersey, New York, Oregon, Pennsylvania, Rhode Island, Vermont, and Washington) have already adopted them or are in the process of doing so.

The 1990 amendments also imposed tighter requirements on certification (an auto's useful life is defined as 100,000 miles instead of the earlier 50,000 miles), on emissions allowed during refueling, on low temperature CO emissions, on in-use performance over time, and on warranties for the most expensive emission control components (8 years/80,000 miles for the catalytic converter, electronic emissions control unit, and onboard emissions diagnostic unit). Regulations were also extended to include nonroad fuels and engines.

Standards for trucks and buses using diesel engines were also strengthened. The 1990 amendments required new urban buses to reduce emissions of diesel particulates 92% by 1996, and all other heavy-duty diesel engines to achieve an 83% reduction by the same year. NO_x emissions must also be reduced, 33% by 1998. Authority to further strengthen these standards led to promulgation in January 2001 of new emission standards requiring a further 90%-95% reduction in emissions phased in over the 2007-2010 model years, and a reduction of 97% in the allowable amount of sulfur in highway diesel fuel. These regulations were followed in May 2004 by similar requirements for nonroad diesel equipment, which are being phased in between 2007 and 2015.

HAZARDOUS AIR POLLUTANTS[4]

Completely rewritten by the Clean Air Act Amendments of 1990, Section 112 of the Act establishes programs for protecting public health and the environment from exposure to toxic air pollutants. As revised by the 1990 amendments, the section contains four major provisions: Maximum Achievable Control Technology (MACT) requirements; health-based standards; standards for stationary "area sources" (small, but numerous sources, such as gas stations or dry cleaners, that collectively emit significant quantities of hazardous pollutants); and requirements for the prevention of catastrophic releases.

First, EPA is to establish technology-based emission standards, called MACT standards, for sources of 187 pollutants listed in the legislation, and to specify categories of sources subject to the emission standards.[5] EPA is to revise the standards periodically (at least every eight years). EPA can, on its own initiative or in response to a petition, add or delete substances or source categories from the lists.

Section 112 establishes a presumption in favor of regulation for the designated chemicals; it requires regulation of a designated pollutant unless EPA or a petitioner is able to show "that there is adequate data on the health and environmental effects of the substance to determine that emissions, ambient concentrations, bioaccumulation or deposition of the substance may not reasonably be anticipated to cause any adverse effects to human health or adverse environmental effects."

EPA is required to set standards for sources of the listed pollutants that achieve "the maximum degree of reduction in emissions" taking into account cost and other non-air-quality factors.

These MACT standards for new sources "shall not be less stringent than the most stringent emissions level that is achieved in practice by the best controlled similar source."

The standards for existing sources may be less stringent than those for new sources, but must be no less stringent than the emission limitations achieved by either the best performing 12% of existing sources (if there are more than 30 such sources in the category or subcategory) or the best performing 5 similar sources (if there are fewer than 30).

Existing sources are given three years following promulgation of standards to achieve compliance, with a possible one-year extension; additional extensions may be available for special circumstances or for certain categories of sources. Existing sources that achieve voluntary early emissions reductions will receive a six-year extension for compliance with MACT.

The second major provision of Section 112 directs EPA to set health-based standards to address situations in which a significant residual risk of adverse health effects or a threat of adverse environmental effects remains after installation of MACT.

This provision requires that EPA, after consultation with the Surgeon General of the United States, submit a report to Congress on the public health significance of residual risks, and recommend legislation regarding such risks. If Congress does not legislate in response to EPA's recommendations, then EPA is required to issue standards for categories of sources of hazardous air pollutants as necessary to protect the public health with an ample margin of safety or to prevent an adverse environmental effect. A residual risk standard is required for any source emitting a cancer-causing pollutant that poses an added risk to the most exposed person of more than one-in-a-million. Residual risk standards are due eight years after promulgation of MACT for the affected source category.

Existing sources have 90 days to comply with a residual risk standard, with a possible two-year extension. In general, residual risk standards do not apply to area sources.

The law directed EPA to contract with the National Academy of Sciences (NAS) for a study of risk assessment methodology, and created a Risk Assessment and Management Commission to investigate and report on policy implications and appropriate uses of risk assessment and risk management. In 1994 NAS published its report, *Science and Judgment in Risk Assessment.* The

Commission study, *Framework for Environmental Health Risk Management*, was released in 1997.

Third, in addition to the technology-based and health-based programs for major sources of hazardous air pollution, EPA is to establish standards for stationary "area sources" determined to present a threat of adverse effects to human health or the environment.

The provision requires EPA to regulate the stationary area sources responsible for 90% of the emissions of the 30 hazardous air pollutants that present the greatest risk to public health in the largest number of urban areas. In setting the standard, EPA can impose less stringent "generally available" control technologies, rather than MACT.

Finally, Section 112 addresses prevention of sudden, catastrophic releases of air toxics by establishing an independent Chemical Safety and Hazard Investigation Board. The Board is responsible for investigating accidents involving releases of hazardous substances, conducting studies, and preparing reports on the handling of toxic materials and measures to reduce the risk of accidents.

EPA is also directed to issue prevention, detection, and correction requirements for catastrophic releases of air toxics by major sources. Section 112(r) requires owners and operators to prepare risk management plans including hazard assessments, measures to prevent releases, and a response program.

NEW SOURCE PERFORMANCE STANDARDS[6]

Section 111 of the Act requires EPA to establish nationally uniform, technology-based standards (called New Source Performance Standards, or NSPS) for categories of new industrial facilities. These standards accomplish two goals: first, they establish a consistent baseline for pollution control that competing firms must meet, and thereby remove any incentive for states or communities to weaken air pollution standards in order to attract polluting industry; and second, they preserve clean air to accommodate future growth, as well as for its own benefits.

NSPS establish maximum emission levels for new major stationary sources—powerplants, steel mills, and smelters, for example—with the emission levels determined by the best "adequately demonstrated" continuous control technology available, taking costs into account. EPA must regularly revise and update NSPS applicable to designated sources as new technology becomes available, since the goal is to prevent new pollution problems from developing and to force the installation of new control technology.

The standards also apply to modifications of existing facilities, through a process called New Source Review (NSR). The law's ambiguity regarding what constitutes a modification (subject to NSR) as opposed to routine maintenance of a facility has led to litigation, with EPA proposing in recent years to modify its interpretation of the requirements of this section.

SOLID WASTE INCINERATORS

Prior to 1990, solid waste incinerators, which emit a wide range of pollutants, were subject to varying degrees of state and federal regulation depending on their size, age, and the type of waste burned. In a new Section 129, the 1990 amendments established more consistent federal requirements specifying that emissions of 10 categories of pollutants be regulated at new and existing incinerators burning municipal solid waste, medical waste, and commercial and industrial waste. The amendments also established emissions monitoring and operator training requirements.

PREVENTION OF SIGNIFICANT DETERIORATION / REGIONAL HAZE

Sections 160-169 of the act establish requirements for the prevention of significant deterioration of air quality (PSD). The PSD program reflects the principle that areas where air quality is better than that required by NAAQS should be protected from significant new air pollution even if NAAQS would not be violated.

The Act divides clean air areas into three classes, and specifies the increments of SO_2 and particulate pollution allowed in each. Class I areas include international and national parks, wilderness and other pristine areas; allowable increments of new pollution are very small. Class II areas include all attainment and not classifiable areas, not designated as Class I; allowable increments of new pollution are modest. Class III represents selected areas that states may designate for development; allowable increments of new pollution are large (but not so large that the area would exceed NAAQS). Through an elaborate hearing and review process, a state can have regions redesignated from Class II to Class III (although none have yet been so redesignated).

While the 1977 amendments only stipulated PSD standards for two pollutants, SO_2 and particulates, EPA is supposed to establish standards for other criteria pollutants. Thus far, only one of the other four (NO_2) has been addressed: the agency promulgated standards for NO_2 in 1988.

Newly constructed polluting sources in PSD areas must install best available control technology (BACT) that may be more strict than that required by NSPS. The justifications of the policy are that it protects air quality, provides an added margin of health protection, preserves clean air for future development, and prevents firms from gaining a competitive edge by "shopping" for clean air to pollute. In Sections 169A and B, the Act also sets a national goal of preventing and remedying impairment of visibility in national parks and wilderness areas, and requires EPA to promulgate regulations to assure reasonable progress toward that goal. In the 1990 Amendments, Congress strengthened these provisions, which had not been implemented.

The amendments required EPA to establish a Grand Canyon Visibility Transport Commission, composed of Governors from each state in the affected region, an EPA designee, and a representative of each of the national parks or wilderness areas in the region. Other visibility transport commissions can be established upon EPA's discretion or upon petition from at least two states. Within 18 months of receiving a report from one of these commissions, EPA is required to promulgate regulations to assure reasonable progress toward the visibility goal, including requirements that states update their State Implementation Plans to contain emission limits, schedules of compliance, and other measures necessary to make reasonable progress. Specifically mentioned is a requirement that states impose Best Available Retrofit Technology on existing sources of emissions impairing visibility.

The Grand Canyon Commission delivered a set of recommendations to EPA in June 1996, and the agency subsequently promulgated a "regional haze" program applicable to all 50 states under this authority.

ACID DEPOSITION CONTROL

The Clean Air Act Amendments of 1990 added an acid deposition control program (Title IV) to the Act. It set goals for the year 2000 of reducing annual SO2 emissions by 10 million tons from 1980 levels and reducing annual NO_x emissions by 2 million tons, also from 1980 levels. The SO2 reductions were imposed in two steps. Under Phase 1, owners/operators of 111 electric generating facilities listed in the law that are larger than 100 megawatts had to meet tonnage emission limitations by January 1, 1995. This would reduce SO_2 emissions by about 3.5 million tons. Phase 2 included facilities larger than 75 megawatts, with a deadline of January 1, 2000. Compliance has been 100%.

To introduce some flexibility in the distribution and timing of reductions, the Act creates a comprehensive permit and emissions allowance system. An allowance is a limited authorization to emit a ton of SO2. Issued by EPA, the allowances would be allocated to Phase 1 and Phase 2 units in accordance with baseline emissions estimates. Powerplants which commence operation after November 15, 1990 would not receive any allowances. These new units would have to obtain allowances (offsets) from holders of existing allowances. Allowances may be traded nationally during either phase. The law also permitted industrial sources and powerplants to sell allowances to utility systems under regulations developed by EPA. Allowances may be banked by a utility for future use or sale.

The Act provided for two types of sales to improve the liquidity of the allowance system and to ensure the availability of allowances for utilities and independent power producers who need them. First, a special reserve fund consisting of 2.8% of Phase 1 and Phase 2 allowance allocations has been set aside for sale. Allowances from this fund (25,000 annually from 1993- 1999 and 50,000 thereafter) are sold at a fixed price of $1,500 an allowance. Independent power producers have guaranteed rights to these allowances under certain conditions. Second, an annual, open auction sold allowances (150,000 from 1993-1995, and 250,000 from 1996-1999) with no minimum price. Utilities with excess allowances may have them auctioned off at this auction, and any person may buy allowances.

The Act essentially caps SO_2 emissions at individual existing sources through a tonnage limitation, and at future plants through the allowance system. First, emissions from most existing sources are capped at a specified emission rate times an historic baseline level. Second, for plants commencing operation after November 15, 1990, emissions must be completely offset with additional reductions at existing facilities beginning after Phase 2 compliance. However, as noted above, the law provides some allowances to future powerplants which meet certain criteria. The utility SO_2 emission cap was set at 8.9 million tons, with some exceptions. The Act provides that if an affected unit does not have sufficient allowances to cover its emissions, it is subject to an excess emission penalty of $2,000 per ton of SO_2 and required to reduce an additional ton of SO_2 the next year for each ton of excess pollutant emitted.

The Act also requires EPA to inventory industrial emissions of SO_2 and to report every five years, beginning in 1995. If the inventory shows that industrial emissions may reach levels above 5.60 million tons per year, then EPA is to take action under the Act to ensure that the 5.60 million ton cap is not exceeded. The Act requires EPA to set specific NO_x emission rate limitations—0.45 lb. per million Btu for tangentially-fired boilers and 0.50 lb. per million Btu for wall-fired boilers—unless those rates can not be achieved by low-NO_x burner technology. Tangentially and wall-fired boilers affected by Phase 1 SO_2 controls must also meet NO_x requirements. EPA was to set emission limitations for other types of boilers by 1997 based on low-NO_x burner costs, which EPA did. In addition, EPA was to propose and promulgate a revised new source performance standard for NO_x from fossil fuel steam generating units, which EPA also did, in 1998. In 2005, EPA used the authority described in the section on "Transported Air Pollution" to further lower the caps on SO_2 and NO_x emissions in the eastern half of the country. These subsequent regulations have been challenged in court and remanded to EPA, but they remain in effect until EPA develops an acceptable replacement. As a result, SO_2 and NO_x emissions have been reduced by a further 50% since 2005.

PERMITS[7]

The Clean Air Act Amendments of 1990 added a Title V to the Act which requires states to administer a comprehensive permit program for the operation of sources emitting air pollutants. These requirements are modeled after similar provisions in the Clean Water Act. Previously, the Clean Air Act contained limited provision for permits, requiring only new or modified major stationary sources to obtain construction permits (under Section 165 of the Act). Sources subject to the permit requirements generally include major sources that emit or have the potential to emit 100 tons per year of any regulated pollutant, plus stationary and area sources that emit or have potential to emit lesser specified amounts of hazardous air pollutants. However, in nonattainment areas, the permit requirements also include sources which emit as little as 50, 25, or 10 tons per year of VOCs, depending on the severity of the region's nonattainment status (serious, severe, or extreme).

States were required to develop permit programs and to submit those programs for EPA approval by November 15, 1993. EPA had one year to approve or disapprove a state's submission in whole or in part. After the effective date of a state plan, sources had 12 months to submit an actual permit application. States are to collect annual fees from sources sufficient to cover the "reasonable costs" of administering the permit program, with revenues to be used to support the agency's air pollution control program. The fee must be at least $25 per ton of regulated pollutants (excluding carbon monoxide). Permitting authorities have discretion not to collect fees on emissions in excess of 4,000 tons per year and may collect other fee amounts, if appropriate. The permit states how much of which air pollutants a source is allowed to emit. As a part of the permit process, a source must prepare a compliance plan and certify compliance. The term of permits is limited to no more than five years; sources are required to renew permits at that time. State permit authorities must notify contiguous states of permit applications that may affect them; the application and any comments of contiguous states must be forwarded to EPA for review. EPA can veto a permit; however, this authority is essentially limited to major permit changes. EPA review need not include permits which simply codify elements of a state's overall clean air plan, and EPA has discretion to not review permits for small sources. Holding a permit to some extent shields a source from enforcement actions: the Act provides that a source cannot be held in violation if it is complying with explicit requirements addressed in a permit, or if the state finds that certain provisions do not apply to that source.

ENFORCEMENT

Section 113 of the Act, which was also strengthened by the 1990 amendments, covers enforcement. The section establishes federal authority to issue agency and court orders requiring compliance and to impose penalties for violations of Act requirements. Section 114 authorizes EPA to require sources to submit reports, monitor emissions, and certify compliance with the Act's requirements, and authorizes EPA personnel to conduct inspections.

Like most federal environmental statutes, the Clean Air Act is enforced primarily by states or local governments; they issue most permits, monitor compliance, and conduct the majority of inspections. The federal government functions as a backstop, with authority to review state actions. The agency may act independently or may file its own enforcement action in cases where it concludes that a state's response was inadequate.

The Act also provides for citizen suits both against persons (including corporations or government agencies) alleged to have violated emissions standards or permit requirements, and against EPA in cases where the Administrator has failed to perform an action that is not discretionary under the Act. Citizen groups have often used the latter provision to compel the Administrator to promulgate regulations required by the statute.

The 1990 Amendments elevated penalties for some knowing violations from misdemeanors to felonies; removed the ability of a source to avoid an enforcement order or civil penalty by ceasing a violation within 60 days of notice; gave authority to EPA to assess administrative penalties; and authorized $10,000 awards to persons supplying information leading to convictions under the Act.

Stratospheric Ozone Protection

Title VI of the 1990 Clean Air Act Amendments represents the United States' primary response on the domestic front to the ozone depletion issue. It also implements the U.S. international responsibilities under the Montreal Protocol on Substances that Deplete the Ozone Layer (and its amendments). Indeed, Section 606(a)(3) provides that the Environmental Protection Agency shall adjust phase-out schedules for ozone-depleting substances in accordance with any future changes in Montreal Protocol schedules. As a result, the phase-out schedules contained in Title VI for various ozone-depleting compounds have now been superseded by subsequent amendments to the Montreal Protocol. Since passage of Title VI, depleting substances such as CFCs, methyl chloroform, carbon tetrachloride, and halons (referred to as Class 1 substances) have been phased out by industrial countries, including the United States. New uses of hydrochlorofluorocarbons (HCFCs) (called Class 2 substances under Title VI) are banned beginning January 1, 2015, unless the HCFCs are recycled, used as a feedstock, or used as a refrigerant for appliances manufactured prior to January 1, 2020. Production of HCFCs is to be frozen January 1, 2015 and phased out by January 1, 2030. Exemptions consistent with the Montreal Protocol are allowed. The EPA is required to add any substance with an ozone depletion potential (ODP) of 0.2 or greater to the list of Class 1 substances and set a phase-out schedule of no more than seven years. For example, methyl bromide (ODP estimated by EPA at 0.7) was added to the list in December 1993, requiring its phaseout by January 1, 2001; this decision was altered by Congress in 1998 to harmonize the U.S. methyl bromide phase-out schedule with the 2005 deadline set by the parties to the Montreal Protocol in 1997. Also, EPA is required to add any substance that is known or may be reasonably anticipated to harm the stratosphere to the list of Class 2 substances and set a phase-out schedule of no more than ten years.

Title VI contains several implementing strategies to avoid releases of ozone-depleting chemicals to the atmosphere, including (1) for Class 1 substances used as refrigerants—lowest achievable level of use and emissions, maximum recycling, and safe disposal required by July 1, 1992; (2) for servicing or disposing refrigeration equipment containing Class 1 and 2 substances—venting banned as of July 1, 1992; (3) for motor vehicle air conditioners containing Class 1 or 2 substances—recycling required by January 1, 1992 (smaller shops by January 1, 1993); (4) sale of small containers of class 1 and 2 substances—banned within two years of enactment; and (5) nonessential products—banned within two years of enactment.

SELECTED REFERENCES

U.S. Environmental Protection Agency. Office of Air Quality Planning and Standards. *Air Trends*. Research Triangle Park, NC. Compiled annually, and available at http://www.epa.gov/airtrends/.

Martineau, Robert J., Jr. and Novello, David P. (eds.). *The Clean Air Act Handbook*. 2nd edition. Chicago: American Bar Association, 2004. 728 p.

For recent issues, see CRS Report R42895, *Clean Air Issues in the 113th Congress: An Overview*, by James E. McCarthy.

Table 4. Major U.S. Code Sections of the Clean Air Act (codified generally as 42 U.S.C. 7401-7671)

42 U.S.C.	Section Title	Clean Air Act, as amended
Subchapter I -	Programs and Activities	
Part A -	Air Quality and Emissions Limitations	
7401	Congressional findings and declaration of purpose	Sec. 101
7402	Cooperative activities	Sec. 102
7403	Research, investigation, training, and other activities	Sec. 103
7404	Research relating to fuels and vehicles	Sec. 104
7405	Grants for support of air pollution planning and control programs	Sec. 105
7406	Interstate air quality agencies; program cost limitations	Sec. 106

Table 4. (Continued)

42 U.S.C.	Section Title	Clean Air Act, as amended
7407	Air quality control regions	Sec. 107
7408	Air quality criteria and control techniques	Sec. 108
7409	National primary and secondary ambient air quality standards	Sec. 109
7410	State Implementation Plans for national primary and secondary ambient air quality standards	Sec. 110
7411	Standards of performance for new stationary sources	Sec. 111
7412	Hazardous air pollutants	Sec. 112
7413	Federal enforcement	Sec. 113
7414	Recordkeeping, inspections, monitoring, and entry	Sec. 114
7415	International air pollution	Sec. 115
7416	Retention of state authority	Sec. 116
7417	Advisory committees	Sec. 117
7418	Control of pollution from federal facilities	Sec. 118
7419	Primary nonferrous smelter orders	Sec. 119
7420	Noncompliance penalty	Sec. 120
7421	Consultation	Sec. 121
7422	Listing of certain unregulated pollutants	Sec. 122
7423	Stack heights	Sec. 123
7424	Assurance of adequacy of state plans	Sec. 124
7425	Measures to prevent economic disruption or unemployment	Sec. 125
7426	Interstate pollution abatement	Sec. 126
7427	Public notification	Sec. 127
7428	State boards	Sec. 128
7429	Solid waste combustion	Sec. 129
7430	Emission factors	Sec. 130
7431	Land use authority	Sec. 131
Part B - Ozone Protection (repealed—new provisions related to stratospheric ozone protection are found at 42 U.S.C. 7671 et seq., below)		
Part C - Subpart I -	Prevention of Significant Deterioration of Air Quality Clean Air	
7470	Congressional declaration of purpose	Sec. 160
7471	Plan requirements	Sec. 161
7472	Initial classifications	Sec. 162
7473	Increments and ceilings	Sec. 163
7474	Area redesignation	Sec. 164
7475	Preconstruction requirements	Sec. 165

42 U.S.C.	Section Title	Clean Air Act, as amended
7476	Other pollutants	Sec. 166
7477	Enforcement	Sec. 167
7478	Period before plan approval	Sec. 168
7479	Definitions	Sec. 169
Subpart II -	Visibility Protection	
7491	Visibility protection for federal class I areas	Sec. 169A
7492	Visibility	Sec. 169B
Part D -	Plan Requirements for Nonattainment Areas	
Subpart 1 -	Nonattainment Areas in General	
7501	Definitions	Sec. 171
7502	Nonattainment plan provisions in general	Sec. 172
7503	Permit requirements	Sec. 173
7504	Planning procedures	Sec. 174
7505	Environmental Protection Agency grants	Sec. 175
7505a	Maintenance plans	Sec. 175A
7506	Limitations on certain federal assistance	Sec. 176
7506a	Interstate transport commissions	Sec. 176A
7507	New motor vehicle emission standards in nonattainment areas	Sec. 177
7508	Guidance documents	Sec. 178
7509	Sanctions and consequences of failure to attain	Sec. 179
7509a	International border areas	Sec. 179B
Subpart 2 -	Additional Provisions for Ozone Nonattainment Areas	
7511	Classifications and attainment dates	Sec. 181
7511a	Plan submissions and requirements	Sec. 182
7511b	Federal ozone measures	Sec. 183
7511c	Control of interstate ozone air pollution	Sec. 184
7511d	Enforcement for Severe and Extreme ozone nonattainment areas for failure to attain	Sec. 185
7511e	Transitional areas	Sec. 185A
7511f	NOx and VOC study	Sec. 185B
Subpart 3 -	Additional Provisions for Carbon Monoxide Nonattainment Areas	
7512	Classification and attainment dates	Sec. 186
7512a	Plan submissions and requirements	Sec. 187
Subpart 4 -	Additional Provisions for Particulate Matter Nonattainment Areas	
7513	Classifications and attainment dates	Sec. 188
7513a	Plan provisions and schedules for plan submissions	Sec. 189
7513b	Issuance of RACM and BACM guidance	Sec. 190

Table 4. (Continued)

42 U.S.C.	Section Title	Clean Air Act, as amended
Subpart 5 -	Additional Provisions for Areas Designated Nonattainment for Sulfur Oxides, Nitrogen Dioxide, or Lead	
7514	Plan submission deadlines	Sec. 191
7514a	Attainment dates	Sec. 192
Subpart 6 -	Savings Provisions	
7515	General savings clause	Sec. 193
Subchapter II -	Emission Standards for Moving Sources	
Part A -	Motor Vehicle Emission and Fuel Standards	
7521	Emission standards for new motor vehicles or new motor vehicle engines	Sec. 202
7522	Prohibited acts	Sec. 203
7523	Actions to restrain violations	Sec. 204
7524	Civil penalties	Sec. 205
7525	Motor vehicle and motor vehicle engine compliance testing and certification	Sec. 206
7541	Compliance by vehicles and engines in actual use	Sec. 207
7542	Information collection	Sec. 208
7543	State standards	Sec. 209
7544	State grants	Sec. 210
7545	Regulation of fuels	Sec. 211
7547	Nonroad engines and vehicles	Sec. 213
7548	Study of particulate emissions from motor vehicles	Sec. 214
7549	High altitude performance adjustments	Sec. 215
7550	Definitions	Sec. 216
7552	Motor vehicle compliance program fees	Sec. 217
7553	Prohibition on production of engines requiring leaded gasoline	Sec. 218
7554	Urban bus standards	Sec. 219
Part B -	Aircraft Emission Standards	
7571	Establishment of standards	Sec. 231
7572	Enforcement of standards	Sec. 232
7573	State standards and controls	Sec. 233
7574	Definitions	Sec. 234
Part C -	Clean Fuel Vehicles	
7581	Definitions	Sec. 241
7582	Requirements applicable to clean-fuel vehicles	Sec. 242
7583	Standards for light-duty clean-fuel vehicles	Sec. 243
7584	Administration and enforcement as per California	Sec. 244

42 U.S.C.	Section Title	Clean Air Act, as amended
	standards	
7585	Standards for heavy-duty clean-fuel vehicles	Sec. 245
7586	Centrally fueled fleets	Sec. 246
7587	Vehicle conversions	Sec. 247
7588	Federal agency fleets	Sec. 248
7589	California pilot test program	Sec. 249
7590	General provisions	Sec. 250
Subchapter III -	General Provisions	
7601	Administration	Sec. 301
7602	Definitions	Sec. 302
7603	Emergency powers	Sec. 303
7604	Citizen suits	Sec. 304
7605	Representation in litigation	Sec. 305
7606	Federal procurement	Sec. 306
7607	Administrative proceedings and judicial review	Sec. 307
7608	Mandatory licensing	Sec. 308
7609	Policy review	Sec. 309
7610	Other authority	Sec. 310
7611	Records and audit	Sec. 311
7612	Economic impact analyses	Sec. 312
7614	Labor standards	Sec. 314
7615	Separability	Sec. 315
7616	Sewage treatment grants	Sec. 316
7617	Economic impact assessment	Sec. 317
7619	Air quality monitoring	Sec. 319
7620	Standardized air quality modeling	Sec. 320
7621	Employment effects	Sec. 321
7622	Employee protection	Sec. 322
7624	Cost of vapor recovery equipment	Sec. 323
7625	Vapor recovery for small business marketers of petroleum products	Sec. 324
7625-1	Exemptions for certain territories	Sec. 325
7625a	Statutory construction	Sec. 326
7626	Authorization of appropriations	Sec. 327
7627	Air pollution from Outer Continental Shelf activities	Sec. 328
Subchapter IV-A	Acid Deposition Control	
7651	Findings and purposes	Sec. 401
7651a	Definitions	Sec. 402
7651b	Sulfur dioxide allowance program for existing and new units	Sec. 403

Table 4. (Continued)

42 U.S.C.	Section Title	Clean Air Act, as amended
7651c	Phase I sulfur dioxide requirements	Sec. 404
7651d	Phase II sulfur dioxide requirements	Sec. 405
7651e	Allowances for states with emissions rates at or below 0.80 lbs./mmBtu	Sec. 406
7651f	Nitrogen oxides emission reduction program	Sec. 407
7651g	Permits and compliance plans	Sec. 408
7651h	Repowered sources	Sec. 409
7651i	Election for additional sources	Sec. 410
7651j	Excess emissions penalty	Sec. 411
7651k	Monitoring, reporting, and recordkeeping requirements	Sec. 412
7651l	General compliance with other provisions	Sec. 413
7651m	Enforcement	Sec. 414
7651n	Clean coal technology regulatory incentives	Sec. 415
7651o	Contingency guarantee, auctions, reserve	Sec. 416
Subchapter V -	Permits	
7661	Definitions	Sec. 501
7661a	Permit programs	Sec. 502
7661b	Permit applications	Sec. 503
7661c	Permit requirements and conditions	Sec. 504
7661d	Notification to administrator and contiguous states	Sec. 505
7661e	Other authorities	Sec. 506
7661f	Small business stationary source technical and environmental compliance assistance program	Sec. 507
Subchapter VI -	Stratospheric Ozone Protection	
7671	Definitions	Sec. 601
7671a	Listing of class I and class II substances	Sec. 602
7671b	Monitoring and reporting requirements	Sec. 603
7671c	Phase-out of production and consumption of class I substances	Sec. 604
7671d	Phase-out of production and consumption of class II substances	Sec. 605
7671e	Accelerated schedule	Sec. 606
7671f	Exchange authority	Sec. 607
7671g	National recycling and emission reduction program	Sec. 608
7671h	Servicing of motor vehicle air conditioners	Sec. 609
7671i	Nonessential products containing chlorofluorocarbons	Sec. 610
7671j	Labeling	Sec. 611
7671k	Safe alternatives policy	Sec. 612

42 U.S.C.	Section Title	Clean Air Act, as amended
7671l	Federal procurement	Sec. 613
7671m	Relationship to other laws	Sec. 614
7671n	Authority of Administrator	Sec. 615
7671o	Transfers among parties to Montreal Protocol	Sec. 616
7671p	International cooperation	Sec. 617
7671q	Miscellaneous provisions	Sec. 618
[29 U.S.C. 655]	Chemical Process Safety Management	Sec. 304 of CAA of 1990
[29 U.S.C. 1662e]	Clean Air Employment Transition Assistance	Sec.1101 of CAA of 1990

Note: This table shows only the major U.S. Code sections. For more detail and to determine when a section was added, the reader should consult the official printed version of the U.S. Code.

End Notes

1 Welfare is defined by the act to include effects on soils, water, crops, vegetation, manmade materials, animals, wildlife, weather, visibility, and climate, damage to and deterioration of property, and hazards to transportation, as well as effects on economic values and on personal comfort and well-being.

2 Unlike the other NAAQS pollutants, ozone is not directly emitted, but rather is formed in the atmosphere by the interaction of volatile organic compounds (VOCs) and nitrogen oxides (NOx) in the presence of sunlight. The control of ozone is, thus, based on regulating emissions of VOCs and NOx.

3 EPA modified the ozone standard, specified in the statute as 0.12 parts per million (ppm) averaged over a one-hour period, to 0.08 ppm averaged over an eight-hour period, through regulations promulgated in July 1997. In April 2004, the agency promulgated an implementation rule for the new eight-hour standard. Under this rule, the one-hour standard was revoked as of June 15, 2005, and areas that had not yet attained it were converted to new classifications depending on their eight-hour concentration of ozone. In general, the former one-hour nonattainment areas remain subject to the controls specified for their one-hour category, but most areas were given additional time to attain the new standard. The ozone standard was revised again in March 2008, to 0.075 ppm averaged over eight hours. Nonattainment areas for the new standard were designated in May 2012, with deadlines to attain the standards extended again.

4 This section of the report was written by Linda-Jo Schierow, Specialist in Environmental Policy.

5 The 1990 amendments specified 189 pollutants, but P.L. 102-187, enacted on December 4, 1991, deleted hydrogen sulfide from the list of toxic pollutants, leaving only 188. On

December 19, 2005, EPA removed methyl ethyl ketone (MEK) from the list of toxic air pollutants. The total number of listed air toxics is now 187.

[6] This section of the report was written by Larry B. Parker, Specialist in Energy and Environmental Policy.

[7] This section of the report was written by Claudia Copeland, Specialist in Environmental Policy.

In: Clean Air and Water
Editors: C. McNulty and R. Huber
ISBN: 978-1-62808-444-3

Chapter 2

SAFE DRINKING WATER ACT (SDWA): A SUMMARY OF THE ACT AND ITS MAJOR REQUIREMENTS*

Mary Tiemann

SUMMARY

This report summarizes the Safe Drinking Water Act (SDWA) and its major programs and regulatory requirements. This report includes the drinking water security provisions added to the SDWA by the Public Health Security and Bioterrorism Preparedness and Response Act of 2002 (P.L. 107-188).

The SDWA, Title XIV of the Public Health Service Act, is the key federal law for protecting public water supplies from harmful contaminants. First enacted in 1974 and substantially amended in 1986 and 1996, the act is administered through programs that establish standards and treatment requirements for public water supplies, control underground injection of wastes, finance infrastructure projects, and protect sources of drinking water. The 1974 law established the current federal-state arrangement in which states may be delegated primary implementation and enforcement authority for the drinking water program. The state-administered Public Water Supply

* This is an edited, reformatted and augmented version of the Congressional Research Service Publication, CRS Report for Congress RL31243, dated October 25, 2012.

Supervision (PWSS) Program remains the basic program for regulating the nation's public water systems, and 49 states have assumed this authority.

The last major reauthorization of the act was done through the Safe Drinking Water Act Amendments of 1996 (P.L. 104-182), which generally authorized appropriations for SDWA programs through FY2003. As with other EPA-administered statutes having expired funding authority, Congress has continued to appropriate funds for the ongoing SDWA programs.

In addition to reviewing key programs and requirements of the SDWA, this report includes statistics on the number and types of regulated public water systems. It also provides tables that list all major amendments, with the year of enactment and public law number, and that cross- reference sections of the act with the major *U.S. Code* sections of the codified statute.

INTRODUCTION

The Safe Drinking Water Act (SDWA), Title XIV of the Public Health Service Act, is the key federal law for protecting public water supplies from harmful contaminants. First enacted in 1974 and substantially amended in 1986 and 1996, the act is administered through programs that establish standards and treatment requirements for public water supplies, control underground injection of wastes, finance infrastructure projects, and protect sources of drinking water. The 1974 law established the current federal-state arrangement in which states may be delegated primary implementation and enforcement authority for the drinking water program. The state- administered Public Water Supply Supervision (PWSS) Program remains the basic program for regulating the nation's public water systems, and 49 states have assumed this authority. The 1996 SDWA amendments generally authorized appropriations for SDWA programs through FY2003. **Table 1** identifies the original enactment and subsequent amendments.

BACKGROUND

As indicated by **Table 1**, the Safe Drinking Water Act has been amended several times since enactment of the Safe Drinking Water Act of 1974 (P.L. 93-523). Congress passed this law after nationwide studies of community water systems revealed widespread water quality problems and health risks resulting from poor operating procedures, inadequate facilities, and uneven management of public water supplies in communities of all sizes. The 1974 law gave the EPA substantial discretionary authority to regulate drinking water contaminants and gave states the lead role in implementation and enforcement.

The first major amendments (P.L. 99-339), enacted in 1986, were largely intended to increase the pace at which the EPA regulated contaminants and to increase the protection of ground water. From 1974 until 1986, the EPA had regulated just one additional contaminant beyond the 22 standards previously developed by the Public Health Service. The 1986 amendments required the EPA to (1) issue regulations for 83 specified contaminants by June 1989 and for 25 more contaminants every three years thereafter, (2) promulgate requirements for disinfection and filtration of public water supplies, (3) ban the use of lead pipes and lead solder in new drinking water systems, (4) establish an elective wellhead protection program around public wells, (5) establish a demonstration grant program for state and local authorities having designated sole- source aquifers to develop groundwater protection programs, and (6) issue rules for monitoring injection wells that inject wastes below a drinking water source. The amendments also increased EPA's enforcement authority.

Congress again amended SDWA with the Lead Contamination Control Act of 1988 (P.L. 100- 572). These provisions were intended to reduce exposure to lead in drinking water by requiring the recall of lead-lined water coolers, and requiring the EPA to issue a guidance document and testing protocol for states to help schools and day care centers identify and correct lead contamination in school drinking water.

Table 1. Safe Drinking Water Act and Amendments (codified generally as 42 U.S.C. 300f-300j)

Year	Act	Public Law Number
1974	Safe Drinking Water Act of 1974	P.L. 93-523
1977	Safe Drinking Water Act Amendments of 1977	P.L. 95-190
1979	Safe Drinking Water Act Amendments	P.L. 96-63
1980	Safe Drinking Water Act Amendments	P.L. 96-502
1986	Safe Drinking Water Act Amendments of 1986	P.L. 99-339
1988	Lead Contamination Control Act of 1988	P.L. 100-572
1996	Safe Drinking Water Act Amendments of 1996	P.L. 104-182
2002	Public Health Security and Bioterrorism Preparedness and Response Act of 2002	P.L. 107-188

After the regulatory schedule mandated in the 1986 amendments proved to be unworkable for the EPA, states, and public water systems, the 104th Congress made sweeping changes to the act with the SDWA Amendments of 1996 (P.L. 104-182). As over-arching themes, the amendments targeted resources to address the greatest health risks, added some regulatory flexibility, provided funding for federal drinking water mandates, and aimed to improve water system compliance capacity. Congress revoked the requirement that the EPA regulate 25 new contaminants every three years and created a risk-based approach for selecting contaminants for regulation. Among other changes, Congress added some flexibility to the standard setting process, required the EPA to conduct health risk reduction and cost analyses for new rules, authorized a drinking water state revolving loan fund (DWSRF) program to help water systems finance projects needed to meet SDWA requirements, added programs to improve small system compliance, expanded consumer information requirements, increased the act's focus on pollution prevention through a voluntary source water protection program, and streamlined the act's enforcement provisions. P.L. 104-182 extended authorizations for appropriations under the act through FY2003.

In 2002, several drinking water security provisions were added to the SDWA through the Public Health Security and Bioterrorism Preparedness and Response Act of 2002 (P.L. 107-188). Title IV of the act included requirements for community water systems serving more than 3,300 individuals to conduct vulnerability assessments and prepare emergency response plans. The law also required the EPA to conduct research on preventing and responding to terrorist or other attacks.

REGULATED PUBLIC WATER SYSTEMS

Federal drinking water regulations apply to the approximately 158,000 privately and publicly owned water systems that provide piped water for human consumption to at least 15 service connections or that regularly serve at least 25 people. Nearly 53,000 of these systems are *community water systems* (CWSs) that serve the same residences year-round. These 53,000 systems provide water to approximately 282 million people. All federal regulations apply to these systems. Another 19,174 public water systems are *non-transient, non-community water systems* (NTNCWS), such as schools or factories, that have their own water supply and generally serve the same individuals for more than six months but not year-round. Most drinking water regulations apply to these systems. More than 86,200 other public water systems are *transient non-community water systems* (TNCWS), such as campgrounds and gas stations, that provide their own water to transitory customers. Only regulations for contaminants that pose immediate health risks apply to these systems.[1]

Most community water systems (85%) are relatively small, serving 3,300 people or fewer; these systems provide water to just 9% of the total population served by community water systems. In contrast, 8% of systems serve populations of 10,000 or more, but provide water to 81% of the population served. Among the community water systems, 77% rely on ground water and 23% rely on surface water. (See Table 2 for statistics on community water systems.)

More than 99% (19,055) of the non-transient, non-community water systems are small or very small and provide water to 83% of the population served by these systems. Approximately 99,700 of the nearly 105,400 non-community water systems (transient and non-transient systems combined) serve 500 or fewer people. These statistics give some insight into the scope of technological, economic, and managerial challenges small public water systems may face in meeting federal drinking water regulations. **Table 3** and **Table 4** provide statistics for non- transient non-community water systems and transient non-community water systems.

Table 2. Size Categories of Community Water Systems

System Size (population served)	Number of Community Water Systems	Percentage of Community Water Systems	Population Served (millions)	Percentage of Population Served
Very small (25-500)	29,666	56%	4.9	2%
Small (501-3,300)	14,389	27%	20.8	7%
Medium (3,301-10,000)	4,748	9%	27.5	10%
Large (10,001-100,000)	3,648	7%	102.7	36%
Very large (>100,000)	386	1%	126.3	45%
Total	54,064	100%	282.3	100%

Source: U.S. Environmental Protection Agency. EPA Safe Drinking Water Information System, *Factoids. Drinking Water and Ground Water Statistics for 2005.* EPA 816-K-03-001. December 2006. Available at http://www.epa.gov/safewater/data/getdata.html.

Table 3. Non-Transient Non-Community Water Systems (NTNCWS)

System Size (population served)	Number of NTNCWS	Percentage of NTNCWS	Population Served (millions)	Percentage of Population Served
Very small (25-500)	16,348	85%	2.28	38%
Small (501-3,300)	2,707	14%	2.71	45%
Medium (3,301-10,000)	102	1%	0.56	9%
Large (10,001-100,000)	17	0%	0.50	8%
Very large (>100,000)	0	0%	0	0%
Total	**19,174**	**100%**	**6.92**	**100%**

Source: EPA Safe Drinking Water Information System, *Factoids. Drinking Water and Ground Water Statistics for 2005.* NTNCWS regularly supplies water to at least 25 of the same people at least six months per year, but not year-round (e.g., schools, factories, office buildings and hospitals which have their own water systems. Most regulatory requirements apply to these systems.

Table 4.Transient Non-Community Water Systems (TNCWS)

System size (population served)	Number of TNCWS	Percentage of TNCWS	Population Served (millions)	Percentage of Population Served
Very small (25-500)	83,351	97%	7.29	52%
Small (501-3,300)	2,721	3%	2.67	19%
Medium (3,301-10,000)	111	0%	0.59	4%
Large (10,001-100,000)	23	0%	0.60	4 %
Very large (>100,000)	4	0%	2.99	21%
Total	**86,210**	**100%**	**14.16**	**100%**

Source: EPA Safe Drinking Water Information System, *Factoids. Drinking Water and Ground Water Statistics for 2005.* Transient non-community water systems provide water in places where people do not remain for long periods of time, such as gas stations and campgrounds. Only regulations for contaminants that pose immediate health risks (e.g. nitrate and bacteria) apply to these systems.

National Drinking Water Regulations

A key component of SDWA is the requirement that the EPA promulgate national primary drinking water regulations for contaminants that may pose health risks and that are likely to be present in public water supplies. Section 1412 instructs the EPA on how to select contaminants for regulation and specifies how the EPA must establish regulations once a contaminant has been selected. The regulations apply to the roughly 168,000 privately and publicly owned water systems that provide piped water for human consumption to at least 15 service connections or that regularly serve at least 25 people. The EPA has issued regulations for more than 90 contaminants, including regulations setting new standards for drinking water disinfectants and their byproducts and for microbial contaminants, a regulation establishing a standard for uranium in drinking water, and a regulation revising the standard for arsenic.

Contaminant Selection and Regulatory Schedules

Section 1412, as amended in 1996, directs the EPA to select contaminants for regulatory consideration based on occurrence, health effects, and meaningful opportunity for health risk reduction. Starting in 1998 and then every five years, the EPA must publish a list of contaminants that may warrant regulation. As of 2001, and every five years thereafter, the EPA must determine whether or not to regulate at least five of the listed contaminants. The act requires the EPA to evaluate contaminants that present the greatest health concern and to regulate those contaminants that occur at concentration levels and frequencies of public health concern. The amendments also included schedules for the EPA to complete regulations for specific contaminants (i.e., radon, arsenic, disinfectants and disinfection byproducts and *Cryptosporidium*).

Standard Setting

For each contaminant that the EPA determines requires regulation, the EPA must set a nonenforceable maximum contaminant level goal (MCLG) at a level at which no known or anticipated adverse health effects occur and which allows an adequate margin of safety. The EPA must then set an enforceable standard, a maximum contaminant level (MCL), as close to the

MCLG as is "feasible" using best technology, treatment techniques, or other means available (taking costs into consideration). The agency generally sets standards based on technologies that are affordable for large communities; however, as amended by P.L. 104-182, the act requires EPA, when issuing a regulation for a contaminant, to list any technologies or other means that comply with the MCL and that are affordable for small public water systems serving populations of 10,000 or fewer. If the EPA does not identify "compliance" technologies that are affordable for small systems, then it must identify small system "variance" technologies or other means that may not achieve the MCL but are protective of public health.

The 1996 amendments authorized the EPA to set a standard at other than the feasible level if the feasible level would lead to an increase in health risks by increasing the concentration of other contaminants or by interfering with the treatment processes used to comply with other SDWA regulations. In such cases, the standard or treatment techniques must minimize the overall health risk. Also, when proposing a regulation, the EPA now must publish a determination as to whether or not the benefits of the standard justify the costs. If the EPA determines that the benefits do not justify the costs, the agency may, with certain exceptions, promulgate a standard that maximizes health risk reduction benefits at a cost that is justified by the benefits.

New regulations generally become effective three years after promulgation. Up to two additional years may be allowed if the EPA (or a state in the case of an individual system) determines the time is needed for capital improvements. (Section 1448 outlines procedures for judicial review of EPA actions involving the establishment of SDWA regulations and other final EPA actions.)

Risk Assessment

The 1996 amendments also added risk assessment and risk communication provisions to SDWA. When developing regulations, the EPA is required to (1) use the best available, peer-reviewed science and supporting studies and data; and (2) make publicly available a risk assessment document that discusses estimated risks, uncertainties, and studies used in the assessment. When proposing drinking water regulations, the EPA must publish a health risk reduction and cost analysis (HRRCA). The EPA may promulgate an interim standard without first preparing this benefit-cost analysis or making a determination as to whether the benefits of a regulation would justify the costs if the Administrator determines that a contaminant presents an urgent threat to public health.

Variances and Exemptions

In anticipation that some systems, particularly smaller ones, could have difficulty complying with every regulation, Congress included in SDWA provisions for variances and exemptions. Section 1415 authorizes a state to grant a public water system a *variance* from a standard if raw water quality prevents meeting the standard despite application of best technology, and the variance does not result in an unreasonable risk to health. A 1996 provision (Subsection 1415(e)) authorizes variances specifically for small systems based on application of best affordable technology. When developing a regulation, if the EPA cannot identify a technology that meets the standard and is affordable for small systems, the EPA must identify variance technologies that are affordable but do not necessarily meet the standard. In cases where the EPA has identified variance technologies, then states may grant small system variances to systems serving 3,300 or fewer persons if the system cannot afford to comply with a standard (through treatment, an alternative water source, or restructuring) and the variance ensures adequate protection of public health. A state may then grant a variance to a small system, allowing the system to use a variance technology to comply with a regulation. With EPA approval, states also may grant variances to systems serving between 3,301 and 10,000 persons. Variances are not available for microbial contaminants. The EPA has determined that affordable compliance technologies are available for all existing standards. Thus, small system variances are not available.

Section 1416 authorizes states to grant public water systems temporary *exemptions* from standards or treatment techniques if a system cannot comply for other compelling reasons (including costs). An exemption is intended to give a water system more time to comply with a regulation and can be issued only if it will not result in an unreasonable health risk. A qualified system may receive an exemption for up to three years beyond the compliance deadline. Systems serving 3,300 or fewer persons may receive a maximum of three additional two-year extensions, for a total exemption duration of nine years.

STATE PRIMACY

Section 1413 authorizes states to assume primary oversight and enforcement responsibility (primacy) for public water systems. To assume primacy, states must adopt regulations at least as stringent as national requirements, develop adequate procedures for enforcement, adopt authority for administrative penalties, maintain records, and develop a plan for providing safe drinking water under emergency circumstances. Currently, 55 of 57 states and territories have primacy authority for the public water system supervision (PWSS) program.

Under Section 1443, Congress authorized appropriations of $100 million annually for the EPA to make grants to states to administer the PWSS program. This section directs the EPA, in accordance with regulations, to allot the sums among the states "on the basis of population, geographical area, number of public water systems, and other relevant factors." States are authorized to use a portion of their drinking water state revolving fund grant (under Section 1452) to cover the costs of administering the PWSS program.

ENFORCEMENT, CONSUMER INFORMATION, AND CITIZEN SUITS

The Safe Drinking Water Act requires public water systems to monitor their water supplies to ensure compliance with drinking water standards and to report monitoring results to the states. States review monitoring data submitted by public water systems, and also conduct their own monitoring, to determine system compliance with drinking water regulations. The EPA monitors public water system compliance primarily by reviewing the violation data submitted by the states.

Section 1414 requires that, whenever the EPA finds that a public water system in a state with primary enforcement authority does not comply with regulations, the agency must notify the state and the system and provide assistance to bring the system into compliance. If the state fails to commence enforcement action within 30 days after the notification, the EPA is authorized to issue an administrative order or commence a civil action. In a nonprimacy state, the EPA must notify an elected local official (if any has jurisdiction over the water system) before commencing an enforcement action against the system.

The 1996 amendments strengthened enforcement authorities, streamlined the process for issuing federal administrative orders, increased administrative penalty amounts, made more sections of the act clearly subject to EPA enforcement, and required states (as a condition of primacy) to have administrative penalty authority. The amendments also provided that no enforcement action may be taken against a public water system that has a plan to consolidate with another system.

Consumer Information and Reports

Enforcement provisions also require public water systems to notify customers of violations of drinking water standards or other requirements, such as monitoring and reporting requirements. Systems must notify customers within 24 hours of any violations that have the potential to cause serious health effects. Additionally, community water systems must mail to all customers an annual "consumer confidence report" on contaminants detected in their drinking water. States are required to prepare annual reports on the compliance of public water systems and to make summaries available to the EPA and the public; the EPA must prepare annual national compliance reports (§1414(c)).

Citizen Suits

Section 1449 provides for citizens' civil actions. Citizen suits may be brought against any person or agency allegedly in violation of provisions of the act, or against the EPA Administrator for alleged failure to perform any action or duty which is not discretionary.

Compliance Improvement Programs

The 1996 amendments added two state-administered programs aimed at improving public water system compliance with drinking water regulations: the operator certification program and the capacity development program. Section 1419 required states to adopt programs for training and certifying operators of community and non-transient non-community systems (e.g., schools and workplaces that have their own wells). The EPA is required to withhold 20% of a state's annual DWSRF grant, unless the state has adopted and implements an operator certification program. Relatedly, Section 1420 required states to establish capacity development programs, also based on EPA guidance. Congress specified that the programs must include (1) legal authority to ensure that new systems have the technical, financial, and managerial capacity to meet SDWA requirements; and (2) a strategy to assist existing systems that are experiencing difficulties to come into compliance. The EPA is required to withhold a portion of SRF grants from states that do not have capacity development strategies. The agency has not had to withhold funds under either of these programs.

GROUND WATER PROTECTION PROGRAMS

Most public water systems rely on ground water as a source of drinking water, and Part C of the act focuses on ground water protection.[2] Section 1421 authorized the establishment of state underground injection control (UIC) programs to protect underground sources of drinking water (USDWs). In 1977, the EPA issued mandated regulations that contained minimum requirements for state UIC programs to prevent underground injection that endangers drinking water sources, and that required states to prohibit any underground injection not authorized by state permit. The law specified that the regulations could not interfere with the underground injection of brine from oil and gas production or recovery of oil unless underground sources of drinking water would be affected.[3]

Section 1422 authorized affected states to submit plans to the EPA for implementing UIC programs and, if approved, to assume primary enforcement responsibility. If a state's plan has not been approved, or the state has chosen not to assume program responsibility, then the EPA must implement the program (§1423). For oil and gas injection operations only, states with UIC programs are delegated primary enforcement authority without meeting EPA regulations under Section 1421, provided states demonstrate that they have an effective program that prevents underground injection that endangers drinking water sources (§1425, added in 1980). EPA has delegated primacy for all classes of wells to 34 states; it shares implementation responsibility in six states, and implements the UIC program for all well classes in 10 states.

To implement this program, EPA placed UIC wells in five classes based on similarity in the fluids injected, and construction, injection depth, design, and operating techniques, and issued regulations that establish performance criteria for each class.[4] In 2008, the agency proposed a rule to establish a new

Class VI well for use in the underground injection of carbon dioxide. Class VI wells would be used for the long-term geologic sequestration of carbon dioxide as a tool for mitigating greenhouse gas emissions from coal-fired power plants and other large stationary sources of carbon dioxide.

A separate groundwater protection provision, Subsection 1424(e), authorizes the EPA to make determinations, on EPA's initiative or upon petition, that an aquifer is the sole or principal drinking water source for an area. In areas that overlie a designated sole-source aquifer, no federal funding may be committed for projects that the EPA determines may contaminate such an aquifer. Any person may petition for sole source aquifer designation. Nationwide, the EPA had designated 73 sole source aquifers.

The act contains three other state programs aimed specifically at protecting ground water. Added in 1986, Section 1427 established procedures for demonstration programs to develop, implement, and assess critical aquifer protection areas already designated by the Administrator as sole source aquifers. Section 1428, also added in 1986, established an elective state program for protecting wellhead areas around public water system wells. If a state established a wellhead protection program by 1989, and the EPA approved the state's program, then the EPA may award grants covering between 50% and 90% of the costs of implementing the program. In 1996, Congress added Section 1429, authorizing the EPA to make 50% grants to states to develop programs to ensure coordinated and comprehensive protection of ground water within the states. For these programs, appropriations were authorized through FY2003 as follows: $15 million per year for Section 1427, $30 million per year for Section 1428, and $15 million per year for Section 1429.

Source Water Assessment and Protection Programs

The 1996 amendments expanded the act's pollution prevention focus to embrace protection of surface water, as well as ground water. Section 1453 required the EPA to publish guidance for states to implement source water assessment programs that delineate boundaries of the areas from which systems receive water, and identify the origins of contaminants in those areas to determine systems' susceptibility to contamination. States with approved assessment programs may adopt alternative monitoring requirements for water systems, as provided for in Section 1418.

Section 1454 authorized a source water petition program based on voluntary partnerships between state and local governments. States may establish a program under which a community water system or local government may submit a petition to the state requesting assistance in developing a voluntary source water quality protection partnership to (1) reduce the presence of contaminants in drinking water, (2) receive financial or technical assistance, and (3) develop a long-term source water protection strategy. This section authorized $5 million each year for grants to states to support petition programs. Also, states may use up to 10% of their DWSRF grant to support various source water protection activities including the petition program.

STATE REVOLVING FUNDS

In 1996, Congress authorized a drinking water state revolving loan fund (DWSRF) program to help systems finance improvements needed to comply with SDWA regulations (§1452). The EPA is authorized to make grants to states to capitalize DWSRFs, which states then may use to make loans to public water systems. States must match 20% of the federal grant. Grants are allotted based on the results of needs surveys. Each state and the District of Columbia must receive at least 1% of the appropriated funds.

Drinking water SRFs may be used to provide loans for expenditures that the EPA has determined will facilitate compliance or significantly further the act's health protection objectives. States must make available 15% of their annual allotment for loan assistance to systems that serve 10,000 or fewer persons, to the extent that funds can be obligated for eligible projects. States may use up to 30% of their DWSRF grant to provide loan subsidies (including forgiveness of principal) to help economically disadvantaged communities. Also, states may use a portion of funds for technical assistance, source water protection and capacity development programs, and for operator certification.

The law authorized appropriations of $599 million for FY1994 and $1 billion per year for FY1995 through FY2003 for DWSRF capitalization grants. The EPA was directed to reserve, from annual DWSRF appropriations, 0.33% for financial assistance to several Trusts and Territories, $10 million for health effects research on drinking water contaminants, $2 million for the costs of monitoring for unregulated contaminants, and up to 2% for technical assistance. The EPA may use 1.5% of funds each year for making grants to Indian Tribes and Alaska Native villages.[5]

DRINKING WATER SECURITY

The 107th Congress passed the Public Health Security and Bioterrorism Preparedness and Response Act of 2002 (P.L. 107-188) to address a wide range of security issues. Title IV of the Bioterrorism Act amended SDWA to address threats to drinking water security. Key provisions are summarized below.

Vulnerability Assessments

Under new SDWA Section 1433, each community water system serving more than 3,300 people was required to conduct an assessment of the system's vulnerability to terrorist attacks or other intentional acts intended to disrupt the provision of a safe and reliable drinking water supply. This provision established deadlines, based on system size, for community water systems to certify to the EPA that they had conducted a vulnerability assessment and to submit to the EPA a copy of the assessment. Section 1433 exempted the contents of the vulnerability assessments from disclosure under the Freedom of Information Act (except for information contained in the certification identifying the system and the date of the certification), and it provides for civil and criminal penalties for inappropriate disclosure of information by government officials.

Section 1433 further required each of these community water systems to prepare an emergency response plan incorporating the results of the vulnerability assessment. The EPA was directed to provide guidance to smaller systems on how to conduct vulnerability assessments, prepare emergency response plans, and address threats.

Section 1433(e) authorized the appropriation of $160 million for FY2002, and such sums as may be necessary for FY2003 through FY2005, to provide financial assistance to community water systems to conduct vulnerability assessments, to prepare response plans, and for expenses and contracts to address basic security enhancements and significant threats.

The Bioterrorism Act also added Sections 1434 and 1435 to SDWA, directing the Administrator to review methods by which terrorists or others could contaminate or otherwise disrupt the provision of safe water supplies. These provisions require the EPA to review methods for preventing, detecting, and responding to such disruptions, and methods for providing alternative drinking water supplies if a water system was destroyed or impaired. Section 1435(e) authorized $15 million for FY2002, and such sums as may be necessary for FY2003 through FY2005 to carry out Sections 1434 and 1435.

Emergency Powers

Under Section 1431, the Administrator has emergency powers to issue orders and commence civil action if (1) a contaminant likely to enter a public drinking water supply system poses a substantial threat to public health, and (2) state or local officials have not taken adequate action. The Bioterrorism Act amended this section to specify that EPA's emergency powers include the authority to act when there is a threatened or potential terrorist attack or other intentional act to disrupt the provision of safe drinking water or to impact the safety of a community's drinking water supply.

Tampering with Public Water Systems

Section 1432 provides for civil and criminal penalties against any person who tampers, attempts to tamper, or makes a threat to tamper with a public water system. Amendments made by the Bioterrorism Act increased criminal and civil penalties for tampering, attempting to tamper, or making threats to tamper with public water supplies. The maximum prison sentence for tampering increased from 5 to 20 years. The maximum prison sentence for attempting to tamper, or making threats to tamper, increased from 3 to 10 years. The maximum fine that may be imposed for tampering increased from $50,000 to $1 million. The maximum fine for attempting to tamper, or threatening to tamper, increased from $20,000 to $100,000.

Emergency Assistance

SDWA Subsection 1442(b) authorizes the EPA to provide technical assistance and to make grants to states and public water systems to assist in responding to and alleviating emergency situations. The Bioterrorism Act amended Subsection 1442(d) to authorize appropriations for such emergency assistance of not more than $35 million for FY2002, and such sums as may be necessary for each fiscal year thereafter.

ADDITIONAL SDWA PROVISIONS

Lead-Free Plumbing

Section 1417 prohibits the use of any pipe, solder, or flux used in the installation or repair of public water systems or plumbing in residential or nonresidential facilities providing drinking water that is not "lead free" (as defined in the act). This section also makes it unlawful to sell pipes, plumbing fittings, or fixtures that are not lead free, or to sell solder or flux that is not lead free (unless it is properly labeled), with the exception of pipes used in manufacturing or industrial processing.[6] Section 1417(e), added in 1996, directed EPA to promulgate regulations setting health-based performance standards limiting the amount of lead that may leach from new plumbing fittings and fixtures, unless a voluntary standard was established within one year of enactment. A voluntary standard was established.

Enacted January 4, 2011, the Reduction of Lead in Drinking Water Act (P.L. 111-380) amended Section 1417 (42 U.S.C. 300g-6) to add new exemptions from prohibitions on the use or sale of lead pipes, solder, and flux; and to revise the SDWA definition of "lead-free." The provisions of this act become effective on January 4, 2014. Briefly, P.L. 111-380 reduced the allowable level of lead in products in contact with drinking water from 8.0% to 0.25% (weighted average), which makes federal law similar to several states' laws, and compatible with new industry standards.

In Section 1417(a), as amended, the general prohibitions on the use of lead pipes, solder, and flux do not apply to "(A) pipes, pipe fittings, plumbing fittings, or fixtures, including backflow preventers, that are used exclusively for nonpotable services such as manufacturing, industrial processing, irrigation, outdoor watering or any other uses where the water is not anticipated to be used for human consumption; or (B) toilets, bidets, urinals, fill valves, flushometer valves, tub fillers, shower valves, service saddles, or water distribution main gate valves that are two inches in diameter or larger."

In Section 1417(d), the term "lead free" means "(A) not containing more than 0.2% lead when used with respect to solder and flux (unchanged from existing law); and (2) not more than a weighted average of 0.25% lead when used with respect to the wetted surfaces of pipes, pipe fittings, plumbing fittings, and fixtures." The legislation established a formula to calculate the weighted average lead content of a pipe, pipe fitting, plumbing fitting, or fixture.

Research, Technical Assistance, and Training

Section 1442 authorizes the EPA to conduct research, studies, and demonstrations related to the causes, treatment, control, and prevention of diseases resulting from contaminants in water. The agency is directed to provide technical assistance to the states and municipalities in administering their public water system regulatory responsibilities. This section authorized $15 million annually for technical assistance to small systems and Indian Tribes, and $25 million for health effects research. (Title II of P.L. 104-182, the 1996 amendments, authorized additional appropriations for drinking water research, not to exceed $26.6 million annually for FY1997 through FY2003.)

Demonstration Grants

The Administrator may make grants to develop and demonstrate new technologies for providing safe drinking water and to investigate health implications involved in the reclamation/reuse of waste waters (§1444).

Records, Inspections, and Monitoring

Section 1445 states that persons subject to requirements under SDWA must establish and maintain records, conduct water monitoring, and provide any information that the Administrator may require by regulation to carry out the requirements of the act. Section 1445(b) authorizes the Administrator or a representative, after notifying the state in writing, to enter and inspect the property of water suppliers or other persons subject to the act's requirements, to determine whether the person is in compliance with the act. Failure to comply with these provisions may result in civil penalties.

This section also directs the EPA to promulgate regulations establishing the criteria for a monitoring program for unregulated contaminants. Beginning in 1999 and every five years thereafter, the EPA must issue a list of not more than 30 unregulated contaminants to be monitored by public water systems. States are permitted to develop representative monitoring plan to assess the occurrence of unregulated contaminants in small systems, and the section authorized $10 million to be appropriated for each of FY1999 through FY2003 to provide grants to cover the costs of monitoring for small systems. All monitoring results are to be included in a national drinking water occurrence data base created under the 1996 amendments.

National Drinking Water Advisory Council

The act established a National Drinking Water Advisory Council, composed of 15 members (with at least two representing rural systems), to advise, consult, and make recommendations to the Administrator on activities and policies derived from the act (§1446).

Federal Agencies

Any federal agency having jurisdiction over federally owned public water systems must comply with all federal, state and local drinking water requirements as well as any underground injection control programs. The act provides for waivers in the interest of national security (§1447).

Assistance to Colonias[7]

Added in 1996, Section 1456 authorized the EPA and other appropriate federal agencies to award grants to Arizona, California, New Mexico, and Texas to provide assistance (not more than 50% of project costs) to *colonias* where the residents are subject to a significant health risk attributable to the lack of access to an adequate and affordable drinking water system. Congress authorized appropriations of $25 million for each of fiscal years 1997 through 1999.

Table 5. *U.S. Code* Sections of the Safe Drinking Water Act (Title XIV of the Public Health Service Act) (42 U.S.C. 300f-300j-26)

42 U.S.C.	Section Title	Safe Drinking Water Act (as amended)
Chapter 6A	**Public Health Service**	
Subchapter XII	**Safety of Public Drinking Water Systems**	
Part A	Definitions	
300f	Definitions	§1401
Part B	**Public Water Systems**	
300g	Coverage	§1411
300g-1	National drinking water regulations	§1412
300g-2	State primary enforcement responsibility	§1413
300g-3	Enforcement of drinking water regulations	§1414
300g-4	Variances	§1415
300g-5	Exemptions	§1416
300g-6	Prohibitions on the use of lead pipes, solder, and flux	§1417
300g-7	Monitoring of contaminants	§1418
300g-8[a]	Operator certification	§1419
300g-9[a]	Capacity development	§1420
Part C	**Protection of Underground Sources of Drinking Water**	
300h	Regulations for state programs	§1421
300h-1	State primary enforcement responsibility	§1422
300h-2	Enforcement of program	§1423
300h-3	Interim regulation of underground injections	§1424
300h-4	Optional demonstration by states relating to oil and natural gas	§1425
300h-5	Regulation of state programs	§1426
300h-6[a]	Sole source aquifer demonstration program	§1427
300h-7[a]	State programs to establish wellhead protection areas	§1428
300h-8[a]	State ground water protection grants	§1429
Part D	**Emergency Powers**	
300i	Emergency powers	§1431
300i-1	Tampering with public water systems	§1432
300i-2[a]	Terrorist and other intentional acts	§1433

42 U.S.C.	Section Title	Safe Drinking Water Act (as amended)
300i-3	Contaminant prevention, detection, and response	§1434
300i-4[a]	Supply disruption prevention, detection and response	§1435
Part E	**General Provisions**	
300j	Assurance of availability of adequate supplies of chemicals necessary for treatment of water	§1441
300j-1[a]	Research, technical assistance, information, training of personnel	§1442
300j-2[a]	Grants for state programs	§1443
300j-3[a]	Special project grants and guaranteed loans	§1444
300j-4[a]	Records and inspections	§1445
300j-5	National Drinking Water Advisory Council	§1446
300j-6	Federal agencies	§1447
300j-7	Judicial reviews	§1448
300j-8	Citizen civil actions	§1449
300j-9	General provisions	§1450
300j-11	Indian Tribes	§1451
300j-12[a]	State revolving loan funds	§1452
300j-13	Source water quality assessment	§1453
300j-14[a]	Source water petition program	§1454
300j-15	Water conservation plan	§1455
300j-16[a]	Assistance to colonias	§1456
300j-17	Estrogenic substances screening program	§1457
300j-18[a]	Drinking water studies	§1458
Part F	**Additional Requirements to Regulate the Safety of Drinking Water**	
300j-21	Definitions	§1461
300j-22	Recall of drinking water coolers with lead-lined tanks	§1462
300j-23	Drinking water coolers containing lead	§1463
300j-24	Lead contamination in school drinking water	§1464
300j-25[a]	Federal assistance for state programs regarding lead contamination in school drinking water	§1465
300j-26	Certification of testing laboratories	[b]

Note: This table shows only the major code sections. For more detail and to determine when a section was added, the reader should consult the official printed version of the *U.S. Code*.

a. These sections include authorizations of appropriations.

b. This provision was added by the Lead Contamination Control Act (P.L. 100-572, §4),which did not amend SDWA.

Estrogenic Substances

Section 1547 authorized the EPA to use the estrogenic substances screening program created in the Food Quality Protection Act of 1996 (P.L. 104-170) to provide for testing of substances that may be found in drinking water if the Administrator determines that a substantial population may be exposed to such substances.

Drinking Water Studies

Section 1458 directed the EPA to conduct drinking water studies involving subpopulations at greater risk and biological mechanisms. EPA also was directed to conduct studies to support specific regulations, including those for disinfectants and disinfection byproducts and *Cryptosporidium*.

SELECTED P.L. 104-182 PROVISIONS NOT AMENDING SDWA

The 104th Congress included a variety of drinking water-related provisions in the 1996 SDWA Amendments that did not amend the Safe Drinking Water Act. Several of these provisions are described below.

Transfer of Funds

Section 302 authorized states to transfer as much as 33% of their annual drinking water state revolving fund grant to the Clean Water Act (CWA) SRF, or an equivalent amount from the CWA SRF to the DWSRF through FY2001. (In subsequent conference reports for EPA appropriations, Congress has authorized states to continue transferring funds.)

Grants to Alaska

Section 303 of the 1996 amendments authorized the EPA to make grants to the State of Alaska to pay 50% of the costs of improving sanitation for rural and Alaska Native villages. Grants are for construction of public water and wastewater systems, and for training and technical assistance programs. Appropriations were authorized at $15 million for each of fiscal years 1997 through 2000. (In P.L. 106-457, Congress reauthorized appropriations for these rural sanitation grants at a level of $40 million for each of fiscal years 2001 through 2005.)

Bottled Water

Section 305 revised Section 410 of the Federal Food, Drug, and Cosmetic Act to require the Secretary of Health and Human Services to issue bottled drinking water standards for contaminants regulated under SDWA within 180 days after the EPA promulgates the new standards, unless the Secretary determines that a standard is not necessary.

Wastewater Assistance to Colonias

Section 307 authorized EPA to make grants to colonias for wastewater treatment works. Appropriations were authorized at $25 million for each of fiscal years 1997 through 1999.

Additional Infrastructure Funding

Section 401 authorized additional assistance, up to $50 million for each of fiscal years 1997 through 2003, for a grant program for infrastructure and watershed protection projects.

Additional Reading

U.S. Government Accountability Office. *EPA Should Improve Implementation of Requirements on Whether to Regulate Additional Contaminants*. GAO-11-254. May 27, 2011. 146. http://www.gao.gov/ products/GAO-11-254.

U.S. Congressional Budget Office. *Public Spending on Transportation and Water Infrastructure*. Pub. No. 4088. November 2010. 49.

National Research Council. *Health Implications of Perchlorate Ingestion.* Board on Environmental Studies and Toxicology. National Academies Press. January 2005. 177.

U.S. Environmental Protection Agency. Office of Water. *Drinking Water Infrastructure Needs Survey and Assessment: Fourth Report to Congress.* EPA 816-R-09-001. March 2009. http://www.epa.gov/safewater/needs.html.

U.S. Environmental Protection Agency. Office of Water. *Drinking Water State Revolving Fund Program: Increasing Impact, 2009 Annual Report.* Office of Water. Report No. EPA 816-R-010-021. November 2010. 44 p. http://water

U.S. Environmental Protection Agency. Office of Water. *Providing Safe Drinking Water in America: 2009 National Public Water Systems Compliance Report.* Office of Enforcement and Compliance Assurance. EPA document No. 305R11001. May 16, 2011. 92.

End Notes

[1] EPA's long-standing policy is to exclude transient systems from drinking water regulations except for those contaminants, such as nitrate, that the EPA believes have the potential to cause immediate adverse human health effects resulting from short-term exposure. (Source: Environmental Protection Agency. National Primary Drinking Water Regulation on Lead and Copper, minor revisions. January 12, 2000 (65 *Federal Register* 1950).)

[2] According to the EPA, of roughly 156,600 public water systems, 142,400 rely on ground water and 14,200 rely on surface water. Among roughly 51,700 community water systems, 40,000 rely on ground water and 11,700 rely on surface water.

[3] The Energy Policy Act of 2005 (P.L. 109-58, §322) amended SDWA Section 1421(d) to specify that the definition of "underground injection" excludes the injection of fluids or propping agents (other than diesel fuels) used in hydraulic fracturing operations related to oil, gas, or geothermal production activities.

[4] The wells are classified as follows: Class I (inject hazardous wastes, industrial non-hazardous liquids, or municipal wastewater beneath the lowermost USDW); Class II (inject brines and other fluids associated with oil and gas production, and hydrocarbons for storage); Class III (inject fluids associated with solution mining of minerals beneath the lowermost USDW); Class IV (inject hazardous or radioactive wastes into or above USDWs and generally are banned); and Class V (all other injection wells—many of these wells inject non-hazardous fluids into or above USDWs and are typically shallow, on-site disposal systems).

[5] For more information, CRS Report RS22037, *Drinking Water State Revolving Fund (DWSRF): Program Overview and Issues*, by Mary Tiemann.

[6] Section 1417(d) defines "lead free" to mean not more than 0.2% lead for solders and fluxes, and not more than 8% lead for pipes and pipe fittings.

[7] Colonias generally are described as unincorporated communities or housing developments on the U.S. side of the U.S.-Mexico border that lack some or all basic infrastructure including plumbing and public water and sewer.

In: Clean Air and Water
Editors: C. McNulty and R. Huber
ISBN: 978-1-62808-444-3

Chapter 3

CLEAN WATER ACT: A SUMMARY OF THE LAW*

Claudia Copeland

SUMMARY

The principal law governing pollution of the nation's surface waters is the Federal Water Pollution Control Act, or Clean Water Act. Originally enacted in 1948, it was totally revised by amendments in 1972 that gave the act its current dimensions. The 1972 legislation spelled out ambitious programs for water quality improvement that have since been expanded and are still being implemented by industries and municipalities.

This report presents a summary of the law, describing the statute without discussing its implementation.

The Clean Water Act consists of two major parts, one being the provisions which authorize federal financial assistance for municipal sewage treatment plant construction. The other is the regulatory requirements that apply to industrial and municipal dischargers. The act has been termed a technology-forcing statute because of the rigorous demands placed on those who are regulated by it to achieve higher and higher levels of pollution abatement under deadlines specified in the law. Early on, emphasis was on controlling discharges of conventional

* This is an edited, reformatted and augmented version of a Congressional Research Service publication, CRS Report for Congress RL30030, from www.crs.gov, prepared for Members and Committees of Congress, dated November 30, 2012.

pollutants (e.g., suspended solids or bacteria that are biodegradable and occur naturally in the aquatic environment), while control of toxic pollutant discharges has been a key focus of water quality programs more recently.

Prior to 1987, programs were primarily directed at point source pollution, that is, wastes discharged from discrete sources such as pipes and outfalls. Amendments to the law in that year authorized measures to address nonpoint source pollution (runoff from farm lands, forests, construction sites, and urban areas), which is estimated to represent more than 50% of the nation's remaining water pollution problems. The act also prohibits discharge of oil and hazardous substances into U.S. waters.

Under this act, federal jurisdiction is broad, particularly regarding establishment of national standards or effluent limitations. Certain responsibilities are delegated to the states, and the act embodies a philosophy of federal-state partnership in which the federal government sets the agenda and standards for pollution abatement, while states carry out day-to-day activities of implementation and enforcement.

To achieve its objectives, the act is based on the concept that all discharges into the nation's waters are unlawful, unless specifically authorized by a permit, which is the act's principal enforcement tool. The law has civil, criminal, and administrative enforcement provisions and also permits citizen suit enforcement.

Financial assistance for constructing municipal sewage treatment plants and certain other types of water quality improvements projects is authorized under title VI. It authorizes grants to capitalize State Water Pollution Control Revolving Funds, or loan programs. States contribute matching funds, and under the revolving loan fund concept, monies used for wastewater treatment construction are repaid to states, to be available for future construction in other communities.

INTRODUCTION

The principal law governing pollution of the nation's surface waters is the Federal Water Pollution Control Act, or Clean Water Act. Originally enacted in 1948, it was totally revised by amendments in 1972 that gave the act its current shape. The 1972 legislation spelled out ambitious programs for water quality improvement that have since been expanded and are still being implemented by industries, municipalities and others. Congress made fine-tuning amendments in 1977, revised portions of the law in 1981, and enacted further amendments in 1987.

This report presents a summary of the law, describing the statute. Many details and secondary provisions are omitted here, and even some major components are only briefly mentioned. Further, this report describes the statute, while other CRS products discuss implementation issues.[1] *Table 1* shows the original enactment and subsequent major amendments. *Table 2*, at the end of this report, cites the major U.S. Code sections of the codified statute.

Table 1. Clean Water Act and Major Amendments (codified generally as 33 U.S.C. §§1251-1387)

Year	Act	Public Law
1948	Federal Water Pollution Control Act	P.L. 80-845
		(Act of June 30, 1948)
1956	Water Pollution Control Act of 1956	P.L. 84-660 (Act of July 9, 1956)
1961	Federal Water Pollution Control Act Amendments	P.L. 87-88
1965	Water Quality Act of 1965	P.L. 89-234
1966	Clean Water Restoration Act	P.L. 89-753
1970	Water Quality Improvement Act of 1970	P.L. 91-224, Part I
1972	Federal Water Pollution Control Act Amendments	P.L. 92-500
1977	Clean Water Act of 1977	P.L. 95-217
1981	Municipal Wastewater Treatment Construction Grants Amendments	P.L. 97-117
1987	Water Quality Act of 1987	P.L. 100-4

Authorizations for appropriations to support the law generally expired at the end of FY1990 (September 30, 1990). Programs did not lapse, however, and Congress has continued to appropriate funds to carry out the act. Since the 1987 amendments, although Congress has enacted several bills that reauthorize and modify a number of individual provisions in the law, none comprehensively addressed major programs or requirements.

BACKGROUND

The Federal Water Pollution Control Act of 1948 was the first comprehensive statement of federal interest in clean water programs, and it specifically provided state and local governments with technical assistance funds to address water pollution problems, including research. Water pollution was viewed as primarily a state and local problem, hence, there were no federally required goals, objectives, limits, or even guidelines. When it came to enforcement, federal involvement was strictly limited to matters involving interstate waters and only with the consent of the state in which the pollution originated.

During the latter half of the 1950s and well into the 1960s, water pollution control programs were shaped by four laws that amended the 1948 statute. They dealt largely with federal assistance to municipal dischargers and with federal enforcement programs for all dischargers. During this period, the federal role and federal jurisdiction were gradually extended to include navigable intrastate, as well as interstate, waters. Water quality standards became a feature of the law in 1965, requiring states to set standards for interstate waters that would be used to determine actual pollution levels and control requirements. By the late 1960s, there was a widespread perception that existing enforcement procedures were too time-consuming and that the water quality standards approach was flawed because of difficulties in linking a particular discharger to violations of stream quality standards. Additionally, there was mounting frustration over the slow pace of pollution cleanup efforts and a suspicion that control technologies were being developed but not applied to the problems. These perceptions and frustrations, along with increased public interest in environmental protection, set the stage for the 1972 amendments.

The 1972 statute did not continue the basic components of previous laws as much as it set up new ones. It set optimistic and ambitious goals, required all municipal and industrial wastewater to be treated before being discharged into waterways, increased federal assistance for municipal treatment plant construction, strengthened and streamlined enforcement, and expanded the federal role while retaining the responsibility of states for day-to-day implementation of the law.

The 1972 legislation declared as its objective the restoration and maintenance of the chemical, physical, and biological integrity of the nation's waters. Two goals also were established: zero discharge of pollutants by 1985 and, as an interim goal and where possible, water quality that is both "fishable" and "swimmable" by mid-1983. While those dates have passed, the goals remain, and efforts to attain them continue.

OVERVIEW

The Clean Water Act (CWA) today consists of two parts, broadly speaking, one being the Title II and Title VI provisions, which authorize federal financial assistance for municipal sewage treatment plant construction. The other is regulatory requirements, found throughout the act, that apply to industrial and municipal dischargers.

The act has been termed a technology-forcing statute because of the rigorous demands placed on those who are regulated by it to achieve higher and higher levels of pollution abatement. Industries were given until July 1, 1977, to install "best practicable control technology" (BPT) to clean up waste discharges. Municipal wastewater treatment plants were required to meet an equivalent goal, termed "secondary treatment," by that date. (Municipalities unable to achieve secondary treatment by that date were allowed to apply for case-by-case extensions up to July 1, 1988. According to EPA, 86% of all cities met the 1988 deadline; the remainder were put under administrative or court-ordered schedules requiring compliance as soon as possible. However, many cities continue to make investments in building or upgrading facilities needed to achieve secondary treatment, and funding needs remain high; see discussion below.) Cities that discharge wastes into marine waters were eligible for case-by-case waivers of the secondary treatment requirement, where sufficient showing could be made that natural factors provide significant elimination of traditional forms of pollution and that both balanced populations of fish, shellfish, and wildlife and water quality standards would be protected.

The primary focus of BPT was on controlling discharges of conventional pollutants, such as suspended solids, biochemical oxygen demanding materials, fecal coliform and bacteria, and pH. These pollutants are substances that are biodegradable (i.e., bacteria can break them down), occur naturally in the aquatic environment, and deplete the dissolved oxygen concentration in water, which is necessary for fish and other aquatic life.

The act also mandated greater pollutant cleanup than BPT by no later than March 31, 1989, generally requiring that industry use the "best available technology" (BAT) that is economically achievable. BAT level controls generally focus on toxic substances. Compliance extensions of as long as two years are available for industrial sources utilizing innovative or alternative technology. Failure to meet statutory deadlines could lead to enforcement action.

The CWA utilizes both water quality standards and technology-based effluent limitations to protect water quality. Technology-based effluent limitations are specific numerical limitations established by EPA and placed on certain pollutants from certain sources. They are applied to industrial and municipal sources through numerical effluent limitations in discharge permits issued by states or EPA (see discussion of "Permits, Regulations, and Enforcement," below). Water quality standards are standards for the overall quality of water. They consist of the designated beneficial use or uses of a waterbody (recreation, water supply, industrial, or other), plus a numerical or narrative statement identifying maximum concentrations of various pollutants that would not interfere with the designated use. The act requires each state to establish water quality standards for all bodies of water in the state. These standards serve as the backup to federally set technology-based requirements by indicating where additional pollutant controls are needed to achieve the overall goals of the act. In waters where industrial and municipal sources have achieved technology-based effluent limitations, yet water quality standards have not been met, dischargers may be required to meet additional pollution control requirements. For each of these waters, the act requires states to set a total maximum daily load (TMDL) of pollutants at a level that ensures that applicable water quality standards can be attained and maintained. A TMDL is both a planning process for attaining water quality standards and a quantitative assessment of pollution problems, sources, and pollutant reductions needed to restore and protect a river, stream, or lake. Based on state reports, EPA estimates that more than 41,000 U.S. waters are impaired and require preparation of TMDLs.

Control of toxic pollutant discharges has been a key focus of water quality programs. In addition to the BPT and BAT national standards, states are required to implement control strategies for waters expected to remain polluted by toxic chemicals even after industrial dischargers have installed the best available cleanup technologies required under the law. Development of management programs for these post-BAT pollutant problems was a prominent element in the 1987 amendments and is a key continuing aspect of CWA implementation.

Prior to the 1987 amendments, programs in the Clean Water Act were primarily directed at point source pollution, wastes discharged from discrete and identifiable sources, such as pipes and other outfalls. In contrast, except for general planning activities, little attention had been given to nonpoint source pollution (runoff of stormwater or snowmelt from agricultural lands, forests, construction sites, and urban areas), despite estimates that it represents more than 50% of the nation's remaining water pollution problems. As it travels across land surface towards rivers and streams, rainfall and snowmelt runoff picks up pollutants, including sediments, toxic materials, and conventional wastes (e.g., nutrients) that can degrade water quality.

The 1987 amendments authorized measures to address such pollution by directing states to develop and implement nonpoint pollution management programs (Section 319 of the act). States were encouraged to pursue groundwater protection activities as part of their overall nonpoint pollution control efforts. Federal financial assistance was authorized to support demonstration projects and actual control activities. These grants may cover up to 60% of program implementation costs.

The CWA provides for special regulation of the discharge of oil or hazardous substances, because of the potentially catastrophic effects of such events on public health and welfare. Section 311 prohibits the discharge of oil or hazardous substances into U.S. waters. It also requires higher standards of care in the management and movement of oil, including a requirement for spill prevention plans; it enables the government to recover the costs of cleaning up oil and hazardous substance discharges; and it provides for penalties for such discharges. In 1990, Congress enacted the Oil Pollution Act, which partially amended Section 311 and established a comprehensive system for the cleanup of oil spills, adding a mechanism to impose liability for such spills.[2]

While the act imposes great technological demands, it also recognizes the need for comprehensive research on water quality problems. This is provided throughout the statute, on topics including pollution in the Great Lakes and Chesapeake Bay, in-place toxic pollutants in harbors and navigable waterways, and water pollution resulting from mine drainage. The act also authorizes support to train personnel who operate and maintain wastewater treatment facilities.

Federal and State Responsibilities

Under this act, federal jurisdiction is broad, particularly regarding establishment of national standards or effluent limitations. The EPA issues regulations containing the BPT and BAT effluent standards applicable to categories of industrial sources (such as iron and steel manufacturing, organic chemical manufacturing, petroleum refining, and others). Certain responsibilities can be assumed by qualified states, in lieu of EPA, and this act, like other environmental laws, embodies a philosophy of federal-state partnership in which the federal government sets the agenda and standards for pollution abatement, while states carry out day-to-day activities of implementation and enforcement. Responsibilities under the act that may be carried out by qualified states include authority to issue discharge permits to industries and municipalities and to enforce permits; 46 states have been authorized to administer this permit program. EPA issues discharge permits in the remaining states—Idaho, Massachusetts, New Hampshire, New Mexico—the District of Columbia, and most U.S. territories. In addition, as noted above, all states are responsible for establishing water quality standards.

Titles II and VI—Municipal Wastewater Treatment Construction

Federal law has authorized grants for planning, design, and construction of municipal sewage treatment facilities since 1956 (Act of July 9, 1956, or P.L. 84-660). Congress greatly expanded this grant program in 1972 in order to assist cities in meeting the act's pollution control requirements. Since that time Congress has authorized $65 billion and appropriated more than $88 billion in CWA funds to aid wastewater treatment plant construction. Grants are allocated among the states according to a complex statutory formula that combines two factors: state population and an estimate of municipal sewage treatment funding needs derived from a survey conducted periodically by EPA and the states. The most recent estimate indicated that, as of January 2008, $298 billion more would be required to build and upgrade municipal wastewater treatment plants in the United States and for other types of water quality improvement projects that are eligible for funding under the act, a 17% increase over the previous estimate from four years earlier.

Under the Title II construction grants program established in 1972, federal grants were made for several types of projects based on a priority list established by the states. Grants were generally available for as much as 55% of total project costs. For projects using innovative or alternative technology (such as reuse or recycling of water), as much as 75% federal funding was allowed. Recipients were responsible for non-federal costs but were not required to repay federal grants.

Policymakers have debated the balance between assisting municipal funding needs, which remain large, and the impact of aid programs such as the Clean Water Act's on federal spending and budget deficits. In the 1987 amendments, Congress balanced these competing priorities by extending federal aid for wastewater treatment construction through FY1994, yet providing a transition towards full state and local government responsibility for financing after that date. Grants under the previous Title II program were authorized through FY1990. Under Title VI of the act, grants to capitalize State Water Pollution Control Revolving Funds, or loan programs, were authorized beginning in FY1989 to replace the Title II grants. States contribute matching funds, and under the revolving loan fund concept, monies used for wastewater treatment construction will be repaid to a state fund, to be available for future construction in other communities.

All states now have functioning loan programs, but the shift from federal grants to loans has been easier for some than others. The new financing requirements have been a problem for cities (especially small towns) that have difficulty repaying project loans. Statutory authorization for grants to capitalize state loan programs expired in 1994; however, Congress has continued to provide annual appropriations. An issue affecting some cities is overflow discharges of inadequately treated wastes from municipal sewers and how cities will pay for costly remediation projects. In 2000, Congress amended the act to authorize a two-year $1.5 billion grant program to help cities reduce these wet weather flows. Authorization for that wet weather grant program expired at the end of FY2003 and has not been renewed.

PERMITS, REGULATIONS, AND ENFORCEMENT

To achieve its objectives, the CWA embodies the concept that all discharges into the nation's waters are unlawful, unless specifically authorized by a permit. Thus, more than 65,000 conventional industrial and municipal dischargers must obtain permits from EPA (or qualified states) under the act's National Pollutant Discharge Elimination System (NPDES) program (authorized in Section 402 of the act). NPDES permits also are required for more than 150,000 industrial and municipal sources of stormwater discharges. An NPDES permit requires the discharger (source) to attain technology-based effluent limits (BPT or BAT for industry, secondary treatment for municipalities, or more stringent where needed for water quality protection). Permits specify the effluent limitations a discharger must meet, and the deadline for compliance. Sources also are required to maintain records and to carry out effluent monitoring activities. Permits are issued for no more than five years and must be renewed thereafter to allow continued discharge.

The NPDES permit incorporates numerical effluent limitations issued by EPA. The initial BPT limitations focused on regulating discharges of conventional pollutants, such as bacteria and oxygen-consuming materials. The more stringent BAT limitations emphasize controlling toxic pollutants—

heavy metals, pesticides, and other organic chemicals. In addition to these limitations applicable to categories of industry, EPA has issued water quality criteria for more than 115 pollutants, including 65 named classes or categories of toxic chemicals, or "priority pollutants." These criteria recommend ambient, or overall, concentration levels for the pollutants and provide guidance to states for establishing water quality standards that will achieve the goals of the act.

A separate type of permit is required to dispose of dredged or fill material in the nation's waters, including wetlands. Authorized by Section 404 of the act, this permit program is administered by the U.S. Army Corps of Engineers, subject to and using EPA's environmental guidance. Some types of activities are exempt from permit requirements, including certain farming, ranching, and forestry practices which do not alter the use or character of the land; some construction and maintenance; and activities already regulated by states under other provisions of the act. EPA may delegate certain Section 404 permitting responsibility to qualified states and has done so twice (Michigan and New Jersey). For some time, the act's wetlands permit program has been one of the most controversial parts of the law. Some who wish to develop wetlands maintain that federal regulation intrudes on and impedes private land-use decisions, while environmentalists seek more protection for remaining wetlands and limits on activities that are authorized to take place in wetlands.

Nonpoint sources of pollution, which EPA and states believe are responsible for the majority of water quality impairments in the nation, are not subject to CWA permits or other regulatory requirements under federal law. They are covered by state programs for the management of runoff, under Section 319 of the act.

Other EPA regulations under the CWA include guidelines on using and disposing of sewage sludge and guidelines for discharging pollutants from land-based sources into the ocean. (A related law, the Ocean Dumping Act, 33 U.S.C. §§1401-45, regulates the intentional disposal of wastes into ocean waters.[3]) EPA also provides guidance on technologies that will achieve BPT, BAT, and other effluent limitations.

The NPDES permit, containing effluent limitations on what may be discharged by a source, is the act's principal enforcement tool. EPA may issue a compliance order or bring a civil suit in U.S. district court against persons who violate the terms of a permit. The penalty for such a violation can be as much as $25,000 per day. Stiffer penalties are authorized for criminal violations of the act—for negligent or knowing violations—of as much as $50,000 per day, three years' imprisonment, or both. A fine of as much as

$250,000, 15 years in prison, or both, is authorized for "knowing endangerment"—violations that knowingly place another person in imminent danger of death or serious bodily injury. Finally, EPA is authorized to assess civil penalties administratively for certain well-documented violations of the law. These civil and criminal enforcement provisions are contained in Section 309 of the act. EPA, working with the Army Corps of Engineers, also has responsibility for enforcing against entities who fail to obtain or comply with a Section 404 permit.

While the CWA addresses federal enforcement, the majority of actions taken to enforce the law are undertaken by states, both because states issue the majority of permits to dischargers and because the federal government lacks the resources for day-to-day monitoring and enforcement. Like most other federal environmental laws, CWA enforcement is shared by EPA and states, with states having primary responsibility. However, EPA has oversight of state enforcement and retains the right to bring a direct action where it believes that a state has failed to take timely and appropriate action or where a state or local agency requests EPA involvement. Finally, the federal government acts to enforce against criminal violations of the federal law.

In addition, individuals may bring a citizen suit in U.S. district court against persons who violate a prescribed effluent standard or limitation or permit requirement. Citizens also may bring suit against the Administrator of EPA for failure to carry out a nondiscretionary duty under the act.

SELECTED REFERENCES

Hamilton, Pixie, Timothy Miller, and Donna Myers. "Water Quality in the Nation's Streams and Aquifers—Overview of Selected Findings, 1991-2001." U.S. Geological Survey Circular 1265. May 2004.

Lavelle, Marianne. "Water Woes." U.S. News & World Report. June 4, 2007. pp. 37-53.

U.S. Congressional Budget Office. Public Spending on Transportation and Water Infrastructure. November 2010. Pub. No. 4088. 49 p.

U.S. Environmental Protection Agency. EPA's Draft Report on the Environment, Technical Document. Chapter 2, Purer Water. June 2003. EPA 600-R-03-050. pp. 2-1 - 2-64.

——. Clean Watersheds Needs Survey 2008, Report to Congress. May 2010. EPA-832-R-10-002.

—— Clean Water State Revolving Fund Programs 2009 Annual Report. June 2010. EPA-832-R- 10-001. 36 p.
—— National Water Quality Inventory: Report to Congress for the 2004 Reporting Cycle. January 2009. EPA-841-R-08-001. 37 p.
U.S. Government Accountability Office. Water Quality, A Catalog of Related Federal Programs. GAO/RCED-96-173. June 1996. 64 p.
—— Water Infrastructure: Information on Financing, Capital Planning, and Privatization. GAO-02-764. August 2002. 79 p.

Table 2. Major U.S. Code Sections of the Clean Water Act (codified generally as 33 U.S.C., Chapter 26, Sections 1251-1387)

33 U.S.C.	Section Title	Clean Water Act (as amended)
Subchapter I -	Research and Related Programs	
1251	Declaration of goals and policy	Sec. 101
1252	Comprehensive programs for water pollution control	Sec. 102
1253	Interstate cooperation and uniform laws	Sec. 103
1254	Research, investigations, training and information	Sec. 104
1255	Grants for research and development	Sec. 105
1256	Grants for pollution control programs	Sec. 106
1257	Mine water pollution control demonstrations	Sec. 107
1258	Pollution control in the Great Lakes	Sec. 108
1259	Training grants and contracts	Sec. 109
1260	Applications for training grants or contracts; allocations of grants or contracts	Sec. 110
1261	Award of scholarships	Sec. 111
1262	Definitions and authorizations	Sec. 112
1263	Alaska village demonstration project	Sec. 113
1264	Lake Tahoe study	Sec. 114
1265	In-place toxic pollutants	Sec. 115
1266	Hudson River PCB reclamation demonstration project	Sec. 116
1267	Chesapeake Bay	Sec. 117
1268	Great Lakes	Sec. 118
1269	Long Island Sound	Sec. 119
1270	Lake Champlain Basin program	Sec. 120
1273	Lake Pontchartrain Basin	Sec. 121

33 U.S.C.	Section Title	Clean Water Act (as amended)
1274	Wet weather watershed pilot projects	Sec. 122
Subchapter II -	Grants for Construction of Treatment Works	
1281	Purpose	Sec. 201
1282	Federal share	Sec. 202
1283	Plans, specifications, estimates, and payments	Sec. 203
1284	Limitations and conditions	Sec. 204
1285	Allotment	Sec. 205
1286	Reimbursement and advanced construction	Sec. 206
1287	Authorization	Sec. 207
1288	Areawide waste treatment management	Sec. 208
1289	Basin planning	Sec. 209
1290	Annual survey	Sec. 210
1291	Sewage collection systems	Sec. 211
1292	Definitions	Sec. 212
1293	Loan guarantees for construction of treatment works	Sec. 213
1294	Public information on water recycling, reuse	Sec. 214
1295	Requirements for American materials	Sec. 215
1296	Determination of priority	Sec. 216
1297	Cost-effectiveness guidelines	Sec. 217
1298	Cost effectiveness	Sec. 218
1299	State certification of projects	Sec. 219
1300	Pilot program for alternative water source projects	Sec. 220
1301	Sewer overflow control grants	Sec. 221
Subchapter III -	Standards and Enforcement	
1311	Effluent Limitations	Sec. 301
1312	Water quality related effluent limitations	Sec. 302
1313	Water quality standards and implementation plans	Sec. 303
1314	Information and guidelines	Sec. 304
1315	Water quality inventory	Sec. 305
1316	National standards of performance	Sec. 306
1317	Toxic and pretreatment effluent standards	Sec. 307
1318	Inspections, monitoring, and entry	Sec. 308
1319	Federal enforcement	Sec. 309

Table 2. (Continued)

33 U.S.C.	Section Title	Clean Water Act (as amended)
1320	International pollution abatement	Sec. 310
1321	Oil and hazardous substance liability	Sec. 311
1322	Marine sanitation devices	Sec. 312
1323	Federal facilities pollution control	Sec. 313
1324	Clean lakes	Sec. 314
1325	National study commission	Sec. 315
1326	Thermal discharges	Sec. 316
1327	Financing study	Sec. 317
1328	Aquaculture	Sec. 318
1329	Nonpoint source management programs	Sec. 319
1330	National estuary program	Sec. 320
Subchapter IV -	Permits and Licenses	
1341	Certification	Sec. 401
1342	National pollutant discharge elimination system	Sec. 402
1343	Ocean discharge criteria	Sec. 403
1344	Permits for dredged or fill material	Sec. 404
1345	Disposal of sewage sludge	Sec. 405
1346	Coastal recreation water quality monitoring and notification	Sec. 406
Subchapter V -	General Provisions	
1361	Administration	Sec. 501
1362	General definitions	Sec. 502
1363	Water pollution control advisory board	Sec. 503
1364	Emergency powers	Sec. 504
1365	Citizen suits	Sec. 505
1366	Appearance	Sec. 506
1367	Employee protection	Sec. 507
1368	Federal procurement	Sec. 508
1369	Administrative procedure and judicial review	Sec. 509
1370	State authority	Sec. 510
1371	Other affected authority	Sec. 511
1372	Separability	Sec. 512
1372	Labor standards	Sec. 513
1373	Public health agency coordination	Sec. 514
1374	Effluent standards and water quality information advisory committee	Sec. 515

33 U.S.C.	Section Title	Clean Water Act (as amended)
1375	Reports to Congress	Sec. 516
1376	General authorization	Sec. 517
1377	Indian tribes	Sec. 518
1251 note	Short title	Sec. 519
Subchapter VI -	State Water Pollution Control Revolving Funds	
1381	Grants to states for establishment of revolving funds	Sec. 601
1382	Capitalization grant agreements	Sec. 602
1383	Water pollution control revolving loan funds	Sec. 603
1384	Allotment of funds	Sec. 604
1385	Corrective action	Sec. 605
1386	Audits, reports, and fiscal controls; intended use plan	Sec. 606
1387	Authorization of appropriations	Sec. 607

Note: This table shows only the major code sections. For more detail and to determine when a section was added, the reader should consult the official printed version of the U.S. Code.

End Notes

[1] For example, see CRS Report R41594, Water Quality Issues in the 112th Congress: Oversight and Implementation, by Claudia Copeland.

[2] P.L. 101-380; 33 U.S.C. § 2701 et seq.

[3] CRS Report RS20028, Ocean Dumping Act: A Summary of the Law, by Claudia Copeland.

In: Clean Air and Water
Editors: C. McNulty and R. Huber
ISBN: 978-1-62808-444-3

Chapter 4

OCEAN DUMPING ACT: A SUMMARY OF THE LAW*

Claudia Copeland

SUMMARY

The Marine Protection, Research, and Sanctuaries Act has two basic aims: to regulate intentional ocean disposal of materials, and to authorize related research. Permit and enforcement provisions of the law are often referred to as the Ocean Dumping Act. The basic provisions of the act have remained virtually unchanged since 1972, when it was enacted to establish a comprehensive waste management system to regulate disposal or dumping of all materials into marine waters that are within U.S. jurisdiction, although a number of new authorities have been added. This report presents a summary of the law.

INTRODUCTION

The Marine Protection, Research, and Sanctuaries Act of 1972 (MPRSA, P.L. 92-532) has two basic aims: to regulate intentional ocean disposal of

* This is an edited, reformatted and augmented version of the Congressional Research Service Publication, CRS Report for Congress RS20028, dated November 30, 2012.

materials, and to authorize related research. Title I of the act, which is often referred to as the Ocean Dumping Act, contains permit and enforcement provisions for ocean dumping. Research provisions are contained in Title II; Title IV authorizes a regional marine research program; and Title V addresses coastal water quality monitoring. The third title of the MPRSA, which authorizes the establishment of marine sanctuaries, is not addressed here.

This report presents a summary of the law, describing the essence of the statute. Descriptions of many details and secondary provisions are omitted here, and even some major components are only briefly mentioned. Further, this report describes the statute, but does not discuss its implementation. Table 1 shows the original enactment and subsequent amendments. Table 2, at the end of this report, cites the major U.S. Code sections of the codified statute.

BACKGROUND

The nature of marine pollution requires that it be regulated internationally, since once a pollutant enters marine waters, it knows no boundary. Thus, a series of regional treaties and conventions pertaining to local marine pollution problems and more comprehensive international conventions providing uniform standards to control worldwide marine pollution has evolved over the last 40 years.

At the same time that key international protocols were being adopted and ratified by countries worldwide (in the early 1970s), the United States enacted the MPRSA to regulate disposal of wastes in marine waters that are within U.S. jurisdiction. It implements the requirements of the London Convention, which is the international treaty governing ocean dumping.[1] The MPRSA requires the Environmental Protection Agency (EPA) Administrator, to the extent possible, to apply the standards and criteria binding upon the United States that are stated in the London Dumping Convention. This convention, signed by more than 85 countries, includes annexes that prohibit the dumping of mercury, cadmium, and other substances such as DDT and PCBs, solid wastes and persistent plastics, oil, high-level radioactive wastes, and chemical and biological warfare agents; and requires special permits for other heavy metals, cyanides and fluorides, and medium- and low-level radioactive wastes.

The MPRSA uses a comprehensive and uniform waste management system to regulate disposal or dumping of all materials into ocean waters. Prior to 1972, U.S. marine waters had been used extensively as a convenient alternative to land-based sites for the disposal of various wastes such as sewage sludge, industrial wastes, and pipeline discharges and runoff.

The basic provisions of the act have remained virtually unchanged since 1972, but many new authorities have been added. These newer parts include (1) research responsibilities for EPA; (2) specific direction that EPA phase out the disposal of "harmful" sewage sludges and industrial wastes; (3) a ban on the ocean disposal of sewage sludge and industrial wastes by December 31, 1991; (4) inclusion of Long Island Sound within the purview of the act; and (5) inclusion of medical waste provisions. Authorizations for appropriations to support provisions of the law expired at the end of FY1997 (September 30, 1997). Authorities did not lapse, however, and Congress has continued to appropriate funds to carry out the act.

Table 1. Ocean Dumping Act and Amendments (codified as 33 U.S.C. 1401-1445, 16 U.S.C. 1431-1447f, 33 U.S.C. 2801-2805)

Year	Act	Public Law Number
1972	Marine Protection, Research, and Sanctuaries Act	P.L. 92-532
1974	London Dumping Convention Implementation	P.L. 93-254
1977	Authorization of appropriations	P.L. 95-153
1980	Authorization of appropriations	P.L. 96-381
1980	Authorization of appropriations	P.L. 96-572
1982	Surface Transportation Assistance Act	P.L. 97-424
1986	Budget Reconciliation	P.L. 99-272, §§6061-6065
1986	Water Resources Development Act	P.L. 99-662, §§211, 728, 1172
1987	Water Quality Act of 1987	P.L. 100-4, §508
1988	Ocean dumping research amendments	P.L. 100-627, title I
1988	Ocean Dumping Ban Act	P.L. 100-688, title I
1988	U.S. Public Vessel Medical Waste Anti-Dumping Act of 1988	P.L. 100-688, title III
1990	Regional marine research centers	P.L. 101-593, title III
1992	National Coastal Monitoring Act	P.L. 102-567, title V
1992	Water Resources Development Act	P.L. 102-580, §§504-510

Table 2. Major U.S. Code Sections of the Marine Protection, Research, and Sanctuaries Act (codified as 33 U.S.C. 1401-1445, 16 U.S.C. 1431-1447f, 33 U.S.C. 2801-2805)

	Section Title	Ocean Dumping Act
33 U.S.C.		
1401	Congressional findings, declaration of policy	Sec. 2
1401	Definitions	Sec. 3
	Title I - Permit Program	
1411	Prohibited acts	Sec. 101
1412	Environmental Protection Agency permits	Sec. 102
1413	Corps of Engineers permits	Sec. 103
1414	Permit conditions	Sec. 104
1414a	Special provisions regarding certain dumping sites	Sec. 104A
1414b	Ocean dumping of sewage sludge and industrial waste	Sec. 104B
1414c	Prohibition on disposal of sewage sludge at landfills on Staten Island	Sec. 104C
1415	Penalties	Sec. 105
1416	Relationship to other laws	Sec. 106
1417	Enforcement	Sec. 107
1418	Regulations	Sec. 108
1419	International cooperation	Sec. 109
1420	Authorization of appropriations	Sec. 111
1421	Annual report to Congress	Sec. 112
	Title II - Research Programs	
1441	Monitoring and research programs	Sec. 201
1442	Research on long-term effects	Sec. 202
1443	Research program - ocean dumping and other methods	Sec. 203
1444	Annual reports	Sec. 204
1445	Authorization of appropriations	Sec. 205
	Title III - Marine Sanctuaries	
	Title IV - Regional Marine Research Programs	
16 U.S.C.		
1447	Purposes	Sec. 401
1447a	Definitions	Sec. 402

	Section Title	Ocean Dumping Act
1447b	Regional marine research boards	Sec. 403
1447c	Regional research plans	Sec. 404
1447d	Research grant program	Sec. 405
1447e	Report on research program	Sec. 406
1447f	Authorization of appropriations	Sec. 407
	Title V -National Coastal Monitoring System	
33 U.S.C.		
2801	Purposes	Sec. 501
2802	Definitions	Sec. 502
2803	Comprehensive coastal water quality monitoring program	Sec. 503
2804	Report to Congress	Sec. 504
2805	Authorization of appropriations	Sec. 505

Note: This table shows the major code sections. For more detail and to determine when a section was added, the reader should consult the printed version of the U.S. Code.

Four federal agencies have responsibilities under the Ocean Dumping Act: EPA, the U.S. Army Corps of Engineers, the National Oceanic and Atmospheric Administration (NOAA), and the Coast Guard. EPA has primary authority for regulating ocean disposal of all substances except dredged spoils, which are under the authority of the Corps of Engineers. NOAA is responsible for long-range research on the effects of human-induced changes to the marine environment, while EPA is authorized to carry out research and demonstration activities related to phasing out sewage sludge and industrial waste dumping. The Coast Guard is charged with maintaining surveillance of ocean dumping.

Regulating Ocean Dumping

Title I of the MPRSA prohibits all ocean dumping, except that allowed by permits, in any ocean waters under U.S. jurisdiction, by any U.S. vessel, or by any vessel sailing from a U.S. port. Certain materials, such as high-level radioactive waste, chemical and biological warfare agents, medical waste, sewage sludge, and industrial waste, may not be dumped in the ocean. Permits for dumping of other materials, except dredge spoils, can be issued by the EPA after notice and opportunity for public hearings where the Administrator determines that such dumping will not unreasonably degrade or endanger human health, welfare, the marine environment, ecological systems, or economic potentialities. The law regulates ocean dumping within the area extending 12 nautical miles seaward from the U.S. baseline and regulates transport of material by U.S.- flagged vessels for dumping into ocean waters. EPA designates sites for ocean dumping and specifies in each permit where the material is to be disposed. EPA is required to prepare an annual report of ocean dumping permits for material other than dredged material.[2]

In 1977, Congress amended the act to require that dumping of municipal sewage sludge or industrial wastes that unreasonably degrade the environment cease by December 1981. (However, that deadline was not achieved, and amendments passed in 1988 extended the deadline to December 1991.) In 1986, Congress directed that ocean disposal of all wastes cease at the traditional 12-mile site off the New York/New Jersey coast (that is, it barred issuance of permits at the 12-mile site) and directed that disposal be moved to a new site 106 miles offshore. In 1988, Congress enacted several laws amending the Ocean Dumping Act, with particular emphasis on phasing out sewage sludge and industrial waste disposal in the ocean, which continued despite earlier legislative efforts.

In 1992, Congress amended the act to permit states to adopt ocean dumping standards more stringent than federal standards and to require that permits conform with long-term management plans for designated dumpsites, to ensure that permitted activities are consistent with expected uses of the site.

Virtually all ocean dumping that occurs today is dredged material, sediments removed from the bottom of waterbodies in order to maintain navigation channels and port berthing areas. Other materials that are dumped include vessels, fish wastes, and human remains. The Corps of Engineers issues permits for ocean dumping of dredged material, using EPA's environmental criteria and subject to EPA's concurrence. The bulk of dredged material disposal results from maintenance dredging by the Corps itself or its contractors. According to data compiled by the Corps, each year an average of approximately 40 million cubic yards of dredged sediment material is disposed in the ocean at designated sites.[3] Before sediments can be permitted to be dumped in the ocean, they are evaluated to ensure that the dumping will not cause significant harmful effects to human health or the marine environment. EPA is responsible for developing criteria to ensure that the ocean disposal of dredge spoils does not cause environmental harm. Permits for ocean disposal of dredged material are to be based on the same criteria utilized by EPA under other provisions of the act, and to the extent possible, EPA-recommended dumping sites are used. Where the only feasible disposition of dredged material would violate the dumping criteria, the Corps can request an EPA waiver. Amendments to MPRSA enacted in 1992 expanded EPA's role in permitting of dredged material by authorizing EPA to impose permit conditions or even deny a permit, if necessary to prevent environmental harm to marine waters.

Permits issued under the Ocean Dumping Act specify the type of material to be disposed, the amount to be transported for dumping, the location of the dumpsite, the length of time the permit is valid, and special provisions for surveillance. The EPA Administrator can require a permit applicant to provide information necessary for the review and evaluation of the application.

In addition to issuing individual permits for non-dredged material, EPA has issued general permits under the MPRSA for several types of disposal activities, such as burial at sea of human remains, transportation and disposal of vessels, and disposal of manmade ice piers in Antarctica.

ENFORCEMENT

The act authorizes EPA to assess civil penalties of not more than $50,000 for each violation of a permit or permit requirement, taking into account such factors as gravity of the violation, prior violations, and demonstrations of good faith; however, no penalty can be assessed until after notice and opportunity for a hearing. Criminal penalties (including seizure and forfeiture of vessels) for knowing violations of the act also are authorized. In addition, the act authorizes penalties for ocean dumping of medical wastes (civil penalties up to $125,000 for each violation and criminal penalties up to $250,000, five years in prison, or both). The Coast Guard is directed to conduct surveillance and other appropriate enforcement activities to prevent unlawful transportation of material for dumping, or unlawful dumping. Like many other federal environmental laws, the Ocean Dumping Act allows individuals to bring a citizen suit in U.S. district court against any person, including the United States, for violation of a permit or other prohibition, limitation, or criterion issued under Title I of the act.

In conjunction with the Ocean Dumping Act, the Clean Water Act (CWA) regulates all discharges into navigable waters including the territorial seas. Although these two laws overlap in their coverage of dumping from vessels within the territorial seas, any question of conflict is essentially moot because

EPA has promulgated a uniform set of standards (40 CFR Parts 220- 229). The Ocean Dumping Act preempts the CWA in coastal waters or open oceans, and the CWA controls in estuaries. States are permitted to regulate ocean dumping in waters within their jurisdiction under certain circumstances.

RESEARCH AND COASTAL WATER QUALITY MONITORING

Title II of the MPRSA authorizes two types of research: general research on ocean resources, under the jurisdiction of NOAA; and EPA research related to phasing out ocean disposal activities.

NOAA is directed to carry out a comprehensive, long-term research program on the effects not only of ocean dumping, but also of pollution, overfishing, and other human-induced changes on the marine ecosystem. Additionally, NOAA assesses damages from spills of petroleum and petroleum products.

EPA's research role includes "research, investigations, experiments, training, demonstrations, surveys, and studies" to minimize or end the dumping of sewage sludge and industrial wastes, along with research on alternatives to ocean disposal. Amendments in 1980 required EPA to study technological options for removing heavy metals and certain organic materials from New York City's sewage sludge.

Title IV of the MPRSA established nine regional marine research boards for the purpose of developing comprehensive marine research plans, considering water quality and ecosystem conditions and research and monitoring priorities and objectives in each region. The plans, after approval by NOAA and EPA, are to guide NOAA in awarding research grant funds under this title of the act.

Title V of the MPRSA established a national coastal water quality monitoring program. It directs EPA and NOAA jointly to implement a long-term program to collect and analyze scientific data on the environmental quality of coastal ecosystems, including ambient water quality, health and quality of living resources, sources of environmental degradation, and data on trends. Results of these activities (including intensive monitoring of key coastal waters) are intended to provide information necessary to design and implement effective programs under the Clean Water Act and Coastal Zone Management Act.

End Notes

[1] The 1972 Convention on the Prevention of Marine Pollution by Dumping of Wastes and Other Matters (known as the London Dumping Convention).

[2] See http://www.epa.gov/aboutepa/history/topics/mprsa/04.html.

[3] See http://el.erdc.usace.army.mil/odd/default.htm.

INDEX

D

E

F

G

H

U

V

W